The David & Charles Manual of
ROOFING

John H. Wickersham

David & Charles

To Michèle, Samantha, Justin and Mark

Some children play with wooden bricks;
but you helped me build with real ones.

British Library Cataloguing in Publication Data

Wickersham, John H.
   The David & Charles manual of roofing.
   1. Roofs–Design and construction
   2. Roofs–Maintenance and repair
   I. Title
   695   TH2391

ISBN 0-7153-8698-0

© John H. Wickersham 1987

First published 1987
Second impression 1989
Third impression 1990
Fourth impression 1991
Fifth impression 1992

Typeset by ABM Typographics Limited, Hull
and printed in Great Britain
by Butler & Tanner Limited, Frome and London
for David & Charles plc
Brunel House   Newton Abbot   Devon

Contents

Author's Note   6

Acknowledgements   7

Introduction   9

1 Roofing and the Spare-time Builder   16

2 Recreational and Storage Buildings   30

3 Flat Roofs – Structure and Cover Materials   42

4 Construction of Pitched Structures   63

5 Pitched Roofs – Coverings Compared   87

6 Pitched Roofs – Installing a New Covering   109

7 Ventilation, Insulation and Flashing   153

8 Maintaining and Repairing Roofs   188

9 Designing, Constructing and Repairing
      Rainwater Drainage Systems   206

Appendix 1 Tables   236

Appendix 2 Address List   239

Index   251

Author's Note

There are many ways to build a roof, and even though this manual is concerned solely with roofs on domestic rather than industrial buildings, variations in structure, covering materials, and features of detailing are legion. From the outset I was anxious not to fall into the trap of many amateur builders who confine their writing to case studies of their own self-build homes. Although I have been involved with many building projects, I wanted this to be a comprehensive manual which extends far beyond the experiences of one person. To achieve this, it was essential to receive guidance and technical advice from many experts associated with the roofing industry. Fortunately the project evoked an enthusiastic response, and this unfailing support was the source of much encouragement. In the later acknowledgements, you will note that all chapters have been subjected to careful scrutiny by people with special expertise and qualification. I was privileged to receive hours of their time, and was invited to visit many manufacturers and building schemes throughout Britain.

Far from being a narrow subject, roofing involves many technologies. There is even a sharp subdivision in design, construction and products for pitched as opposed to flat roofs. I have to admit that the first draft of this book was twice as long and contained three times as many illustrations as the finished work. Only a sensible publisher could explain that the manuscript in its entirety would cost far too much to produce. However, when trimming it to size, key issues were retained. If you find a topic which has been missed or has been treated briefly, you should approach the manufacturers, many of whom publish detailed literature on the installation of their products. An address list is provided so that you may seek out this information.

On one hand, this manual of roofing is addressed to home maintenance enthusiasts and amateur builders. It assumes that many of its topics will be put to practical use. However, it also aims to inform the reader who intends calling on the services of a roofing contractor but who wants prior insight into the scope of the project. Thus its approach is both theoretical and practical, and there is no doubt that it will also act as a useful sourcebook for the student of building, or the tradesman. To this end, it intentionally employs a technical vocabulary, albeit with the support of explanatory text and illustration. Whereas it does not claim to be a textbook for the specialist, it is a focus on an aspect of building in which there is a surprising dearth of literature.

Recognising that the chosen topic includes many diverse areas of study, it was acknowledged from the outset that this should be underpinned with textual references to statutory building regulations and the Codes and Standards of the British Standards Institution. But regulations and codes of practice, like the materials to which they refer, are ever-changing. No sooner had the research and writing been completed, when the pending alteration of a particular detail was announced. At first, amendments were awaited and noted. However, it soon became clear that there would always be another change around the corner, and that a manuscript had finally to be printed.

At the time of completion, the British Standard 5268 concerned with trussed rafters was in the process of being amended which meant that sections of text and supporting diagrams had to be altered. Meantime the full implications of revised building regulations for England and Wales were not fully known. Rather than delay publication, it was decided to make reference to the Building Regulations (England and Wales) 1976 with their later amendments with the knowledge that there would be no substantial major changes. The intention of the subsequent statutory documents is to simplify interpretation, to provide less definitive statement, but to preserve the spirit and purpose of their raison d'etre. Suffice it to say, it is incumbent on the reader to check the current situation in respect of all aspects of content, with recognition that building practice, by its very nature, is ever-changing.

With regard to the reader who is contemplating self-build, repairs or replacements two further points must be noted. Firstly, the matter of safety is crucial in any roof level project. Roofing work on most domestic buildings can be straightforward – as long as there is safe access. Secondly, it is important to acknowledge that 'Do-it-Yourself' does not mean attempt it all yourself. Roofing calls on a variety of skills, and whereas some tasks may fall within your grasp, others are best left to professionals. In parts of this manual, you are actually discouraged from attempting certain jobs yourself. Perhaps I should have recognised this strategy when compiling the manual itself. It would certainly have been finished more quickly if I had shared out the tasks of taking the photographs, toiling in a darkroom, completing the drawings, and typing the manuscript.

In conclusion, therefore, take care when climbing ladders and consider the wisdom of entrusting the erection of a scaffold access to a specialist. Icarus showed great ambition when seeking out the heavens, but dodgy equipment and a disregard for his own safety brought him to earth with a bump. If you have no head for heights, be sensible and seek the services of a specialist contractor. This manual will enable you to negotiate the work and verify his workmanship. Otherwise don your overalls, set your sights high, climb with caution, and enjoy working on the roof over your head.

Acknowledgements

To ensure that this manual presents principles of good practice, and contains accurate statements on matters of theory, the text was subjected to close scrutiny by a number of roofing experts. In this quest for accuracy, I am indebted to the following specialists who gave generously of their time and talent:

Tony Huxley, Technical Officer for Weston Hyde Products Ltd checked the data in Chapter 2 relating to roofing structures and Novolux profiled sheeting.

Paul Stark, Market Analyst for Coolag Purlboard Ltd confirmed the accuracy of Chapter 3 in respect of timber structures and flat roofs.

John Muttock, Technical Sales and Services Manager of Permanite Ltd read and verified the accuracy of Chapters 2 & 3. Further assistance was given by David Rotchell, Marketing Manager of Permanite Ltd.

Peter Jackson, Technical Director and John Park, Senior Engineer, of Gang-Nail Ltd confirmed the accuracy of Chapter 4 in respect of trussed rafter roofing, and advised on the implications of BS 5268 Part 3 : 1985. Richard Thomas, Marketing Manager of Gang-Nail Ltd acted as co-ordinator and arranged consent for the reproduction of the illustrations detailed on page 8 from the Company's publication *Trussed Rafter Construction and Specification Guide*.

John Dodd, Product Manager of The Marley Roof Tile Company and Geoff Hawkins, Technical Services Manager of Redland Roof Tiles both read and approved the content of Chapters 5 & 6 with particular reference to concrete tiles. Brian Daniels, Group Publicity Manager of Redland PLC co-ordinated links within the company and arranged for the reproduction of photographs held in the Redland library.

Terry Hughes, Marketing Development Manager, Penrhyn Quarries Ltd, advised and read the manuscript of Chapters 5 & 6 with regard to Welsh slate. Terry also made available the information given in Appendix Table 3 which comes from Penrhyn and Buttermere Slate Roofing Manual.

Mike Wood, Marketing Services Manager of Eternit Building Products Ltd checked the information in Chapters 5 & 6 with regard to fibre cement slates.

Peter Campbell, Architect, Department of the Environment verified the accuracy of Chapters 3 & 7 with regard to insulation, ventilation, and the problem of condensation.

Richard Murdoch, Senior Technical Officer of the Lead Development Association read the manuscripts on lead work which appear in Chapter 7 and made arrangements for the reproduction of drawings itemised on page 8 which are taken from *Lead Sheet in Building*.

David Wickersham, Assistant Divisional Director (Technical Services), Housing Dept, Westminster City Council advised on overall content. Formerly a Building Inspector, David helped with the interpretations of Planning Consents, Building Regulations, and provided information on British Standard Codes of Practice. In the true spirit of brotherhood, he also helped me build the roof on my house.

In addition to this close scrutiny of text, many others contributed towards the compilation of the finished work. I place on record the help of:
Bernard Ballard, Sales Department, Delabole Slate Ltd; Richard Coleman, Inventor of the Jenny Twin slate fixer, RFJ Products; Consumers' Association, 14 Buckingham Street, London, WC2N 6DS; Ken Coulson, Manager Specialist Services, Hydro-Air International (UK) Ltd; Cyril Cripsey, Sales Manager, Mustang Tools Ltd; Peter Dimblebee, Technical Sales Administrator, Eternit Building Products Ltd; Peter Garbett, Secretary, Trussed Rafter & Fabricators' Association; John Gibson, Director, Kestner Building Products Ltd; Arthur Hoare, Senior Scientific Officer, Energy Efficiency Office, Department of Trade and Industry; G. E. Holland, Architectural Sales Liaison Manager, TAC Construction Materials Ltd; Don Howarth of Dalton Howarth, Advertising Agency for Euroroof Ltd; Beryl Kay, Novolux Sales Dept,

Acknowledgements

I.C.I. Hyde Products; Henrietta Lees, Public Relations Executive, The Marley Roof Tile Co; Robin Mackley, The Clay Roofing Tile Council; Valerie Mason, Marketing Services, Ruberoid Building Products Ltd; Neil McEwan, Managing Director, Sumaco Ltd; Judith Mill, Planox Ltd; Anne Peters, Librarian, Timber Research & Development Assoc; John Nancollis, Public Relations Executive, Barry Hook Associates; Mike Newmarch, Group Press Officer, SGB Group PLC; Sarah Norris, Publicity Dept, E. H. Bradley Building Products Ltd; Martin Oldridge, Director, Sandtoft Roofing Tile Manufacturers; Dick Perrott, Former Director, British Lead Manufacturers' Association; Anthony Raynes, Director, J. & D. Raynes & Sons Ltd; Michael Reed, Managing Director, Euroroof Ltd; Andrew Rhodes, Marketing Department, Marley Extrusions Ltd; David Rhodes, Timloc Building Products Ltd; Colin Rozee, Marketing Manager, Willans Building Services; Brian Rust, Marketing Department, Marley Extrusions Ltd; Stephen Shields, Sales Manager, Red Bank Manufacturing Co Ltd; Richard Sheppard, Marketing Manager, Tunnel Building Products, Ltd; John Shillabeer, Director, Cavity Trays Ltd; David Smith, Managing Director, Europalite Ltd; Dean Stiles, Editor, Builders' Merchants News; Russ Swan, Constellation P.R., for O.F.I.C. (GB) Ltd; Andrew Thom, Sales Manager, Ready Scaffolding Ltd; Sally Thompson, Publicity Dept, D. Anderson & Son Ltd; Don Turnbull, Editor, Roofing, Cladding & Insulation, Patey Doyle (Publishing) Ltd; Anthony Vale, Managing Director, Spandoboard Insulations Ltd; Cathy Wells, Marketing Services, Ruberoid Building Products Ltd; Geoffrey Wheeler, Designer, B. C. Barton & Son Ltd; Mary Whitaker, Senior Commercial Assistant, PVC Sheet Division, Western Hyde Products, Ltd; Julia Williams, Assistant Editor, David & Charles Publishers plc; Richard Wormell, General Secretary, The National Association of Roofing Contractors.

Note: Since this manual was completed, several of the people mentioned in these acknowledgements, have taken up new appointments; some manufacturers have also changed their titles. The information above was accurate at the time when the text was written. For the sake of consistency, there has been no attempt to list the technical qualifications of the contributors.

The author would like to thank the following companies for their kind permission to reproduce the photographs listed: Redland Roof Tiles Ltd, title page main picture, Photos 54, 63, 68, 71a-c, 74, 76a-b, 78, 79a-b, 80, 96a-b, 100, 102, 105, 106, 107a-c, 114, 115, 116, 117, 118, 119, 120, 121, 125, 126; The Marley Roof Tile Co Ltd, Title page insert, 4, 62, 65, 66, 69a-e, 70, 75a-e, 77a-c, 81a-c, 83, 109a-b; Willan Building Services, 95, 97, 98, 99; Marley Waterproofing Products, 30, 111, 113; Penrhyn Quarries Ltd, back of jacket, 85, 88, 92c; B. C. Barton & Son Ltd, 9, 10a-d; The Clay Roofing Tile Council, 3, 51; Weston Hyde Products Ltd, 21, 23, 24, 25; Cavity Trays Ltd, 108, 112; O.F.I.C. (GB) Ltd, 22, 26; Burton Wire & Tube Co Ltd, 13; Ready Scaffolding Ltd, 16; Mustang Tools Ltd, 20; Ruberoid Building Products Ltd, 31; Coolag Purlboard Ltd, 33; E. H. Bradley Building Products Ltd, 47; Eternit Building Products Ltd, 57; Tunnel Building Products Ltd, front of jacket, 93; Alumasc Ltd, 131. All other photographs were taken by the author.

The line illustrations on pages 174, 175, 176, 177, 180, 182, and 184 and on the jacket are from *Lead Sheet in Building*, the definitive manual published jointly by the Lead Development Association and the British Lead Manufacturers' Association, designed and illustrated by Donald Dewar-Mills who has generously waived his copyright fee as a gesture of goodwill to the author. The line illustrations on pages 155 and 158 are reproduced with kind permission of the Energy Efficiency Office and come from their publication *Make the Most of your Heating*. The line illustrations on pages 66, 67, 68, 69, 70, 71, 73, 74, 75 and 81 are reproduced with kind permission of Gang-Nail Systems Ltd and come from their publication *Trussed Rafter Construction Specification Guide*. The author would also like to thank the following companies for their permission to reproduce the line illustrations on the pages listed: Gang Nail Ltd, 66, 67, 68, 69, 70, 71, 73, 74, 75, 81; Willan Building Services, 147; The Marley Roof Tile Co Ltd, 166; and Redland Roof Tiles Ltd, 94. Table 3, Appendix 1 page 238 is reproduced courtesy of Penrhyn Quarries Ltd.

John H. Wickersham

Introduction

Roofing – a marriage of beauty and function

The dome of St Paul's was designed to do rather more than keep out the rain. This truism may be applied to any roof irrespective of the building it covers. Whether it is the crowning glory of a Wren cathedral, or just the lid over one's humble home, its requirements remain very similar. At first sight it should be pleasing to the eye; but to be deemed a success, it must fulfil several functional objectives.

There is much you can do to maintain harmony between function and appearance. And whereas this manual deals mostly with practical matters, appearances should not be forgotten. Its focus is the roofs of domestic buildings – houses, home extensions, garages, worksheds and so on. Ordinary roofs on ordinary houses are unlikely to achieve architectural awards, but if you prevent ugly depreciation you are contributing to visual amenity. 'Small' can be beautiful, and, if regularly maintained, the simplest roof enhances the skyline.

In essence, a successful roof represents a marriage of beauty and function. Appearance is important, but so is structural integrity. No matter how magnificent the dome of St Pauls, it would be adjudged a failure if it leaked badly, or posed a danger of collapse.

Photo 1 Ornate detailing with shaped clay tiles and decorative barge boards makes a bold roofline on this country cottage

Introduction

The same can be said for the roof on your home, and as a front-line defence against weather, periodic attention will certainly be necessary. Not long ago, roof repair or replacement was considered the sole preserve of the professional. But today, self-help has passed far beyond the modest DIY enterprise of twenty years ago. People do much more than paper the lounge or pave the patio; the sky is the limit, and many find roof renovation within their grasp.

In a projected new building, roof design and construction are subject to the approval of the local authority. They must also be consulted about intended alterations, and are empowered to disallow a proposal or require that amendments are made. General maintenance work, however, simply rests with the occupier. If a property is rented, it is the tenant rather than the landlord who is liable by law to ensure that certain aspects of roof repair are carried out. Case law defines 'tenantable repair jobs', and replacing a broken tile is an example. However, thanks to the opportunities for hiring access equipment

there is every reason for you to carry out many tasks yourself. The aim of this manual is to provide guidance, explaining both the theoretical 'why' as well as the practical 'how'. Its topics are sometimes simple, but it also includes ambitious undertakings, such as roofing a self-build luxury home.

Before launching into action, however, you should be aware of the many requirements which deem a roof to be successful.

Weather resistance

Being able to resist the weather is not merely desirable but mandatory. With regard to Inner London, the statutory instruments are the London Building (Constructional) By-laws. Different Building Regulations apply to the remainder of England and Wales, although the requirements are much the same. In Scotland the statutory instruments are the Building Standards (Scotland) Regulations 1981. No definition of 'weather' is given, but rain, snow, wind, frost and sun are singly or collectively responsible for causing damage. Wind-borne dust is an associated problem, and older roofs built without a secondary barrier of sarking felt or boarding usually have dirty attics.

Photo 2 Flat roofs are unusual, lacking in character, and prone to problems less evident on the pitched alternative

With regard to wind and snow, the problem is not just a matter of penetration, but loading as well. Snow is heavy stuff, and this has to be acknowledged by the designer of the structure. Likewise, gale-force winds represent a different type of 'loading', and their destructive potential is fully recognised. 'Weather resistance' means much more than coping with a downpour.

Strength and stability

It is not only 'weather loadings' which have to be taken into account by the designer. For instance, a roof space is often designated to house a water tank, and this imposes considerable demand on the structure. An average-sized tank holding 225L (50gal) adds a weight of 225kg (500lb). The weight of covering materials and structural timbers are also acknowledged in the calculations. Carrying a pile of tiles up a scaffolding serves a sharp reminder that the overall covering will be surprisingly heavy. Lastly, the designer will need to ensure that the completed structure is strong enough to withstand the weight of a workman.

A number of factors must therefore be

Photo 3 Plain tiles made in clay, and an eyebrow dormer window suitably enhance a roof which otherwise would have lacked character

taken into account to assure structural stability: dead loads (weight of slates), imposed loads (weight of a maintenance worker) and wind loadings (effect of pressure or suction). These, in turn, will be related by the designer to the spans between load-bearing walls.

Whereas this manual shows that the construction or repair of roofs can be straightforward, it is important to stress that drawing up material specifications, or designing a roof for a residential building are *not* tasks which the amateur builder should attempt.

Durability

A covering material on a roof is exposed to many destructive forces. For example, frost hastens the failure of porous covering materials such as clay tiles. Freezing water which has found its way into the tile can cause it to spall, ie split into small fragments. Exfoliation of outer layers, illustrated in Photo 128 (page 204), is evidence of advanced deterioration. Thermal variation is another contributor to failure, and this is particularly

11

Photo 4 The traditional texture of stone slates is recreated to good effect with Marley Westwold tiles

evident in respect of roofs covered in bitumen felt. The addition of a layer of reflective chippings, which is now mandatory, affords some solar protection on felt-covered flat roofs.

Atmospheric pollution, a growing problem of industrial society, is another test of roofing integrity. Metal details, such as zinc flashing, deteriorate rapidly in certain industrial locations. Salt-laden atmosphere on coastal sites is similarly destructive. In total, the durability of covering materials is put to the test by many destructive influences. If money is unavailable to permit complete replacement, temporary refurbishments are often undertaken using various types of coating. In parts of the West Country a veneer of cement or a proprietory coating is often added to older roofs – particularly if they are situated in an exposed coastal position. No roofing material will last for ever.

Drainage

A roof area acts as a sizeable collector of rainwater, and a system for its collection, transport and dispersal is essential for almost all buildings. Thatched roofs are the exception, and in all but a few cases these are notable for their absence of both guttering and downpipe. Designing a system concerns factors such as pitch, catchment area, siting of downpipes and so on. Planning this provision needs knowledge of flow rates, configuration of gutter runs and the capacities of products being fitted. With the explanations given in Chapter 9, this should prove to be a straightforward paper exercise.

Thermal insulation

A successful roof will contribute towards the preservation of comfortable living temperatures. Retaining heat efficiently and the conservation of energy are topics of our times, and heat loss through poorly insulated roofs has received plenty of publicity. The inclu-

sion of an insulant is mandatory in new properties, and to encourage the upgrading of existing housing, local authorities sometimes offer grant aid.

The matter of heat loss is important, but insulation in the roof space also helps to control solar gain. Most roof coverings are poor reflectors of heat, and in sunny weather the temperature rise in a loft space is considerable. At worst, this may lead to discomfort in the rooms below, but loft insulation will do much to reduce the problem.

Photo 5 Stone slates laid in diminishing courses, and attached by oak pegs, have adorned the gatehouse of Stokesay Castle in Shropshire for several centuries

Improvements in this aspect of building have gained considerable momentum in the last decade. However, the elimination of draughts, combined with the installation of insulants, has brought a new problem – condensation. Concern that many roofs are deteriorating as a result of interstitial condensa-

Introduction

tion has been instrumental in emphasising the urgent need for ventilation.

Ventilation

It is ironical that many householders whose roof space is stuffed with insulant have solved one problem, and introduced another. Coupled with efforts to seal gaps and eliminate draughts, the penalty of these heat-saving strategies has been the formation of condensation in roof voids.

In new buildings, various strategies are adopted to avoid problems, but on your existing property there is much that you can do to effect cures.

Photo 6 Fish scale patterning, though not difficult to create, bestows charm on this single storey dwelling restored in 1840, at Brentwood, Essex

Fire resistance

'Safety in Fire' receives lengthy treatment in the Building Regulations. In respect of roofing, 'penetration' of flame together with the risk of structural collapse are topics of importance. Equally serious is the surface spread of flame which extends damage to adjacent roofs – an acute problem in terrace properties. After the year 1666, lessons learned from London's Great Fire resulted in the introduction of regulations concerning the choice of covering materials. Wooden shingles and thatch pose special risks, and the need to introduce legislative measures regarding their use was recognised several centuries ago. In respect of more modern materials, corrugated plastic sheeting used to present a fire hazard, but self-extinguishing PVC roofing sheets are now manufactured.

Appearance

Appearance is important and a good roof should be in keeping with the rural or urban landscape. Local character in building style has evolved throughout history, and regional differences in the availability of building materials has established distinctive patterns. Reed thatch in Norfolk, Swithland slate in Leicestershire, sandstone slabs in Sussex and stone roofs in the Cotswolds are examples of coverings which bestow much-loved character on our national heritage. Preserving regional variation is a shared duty of planners, architects, developers and ourselves as occupants.

Other functions

Inevitably there are aspects of roofing which have been omitted from this manual. For

Photo 7 Thatching is an unsuitable material in urban settings, although this decorative roof in Leighton Buzzard sits comfortably on a timber frame building

example, houses in the vicinity of major airports require special measures to attenuate aircraft noise; but the construction of acoustical barriers in roof voids is a subject not covered. Equally specialised is the construction of roofs providing natural-light sources. Factories, sports halls and public auditoria often have roof structures which feature domes or panels of glare-diffusing membranes. Although PVC roof sheets are discussed, the subject of roof lights has not been included.

This list shows that there is rather more to roofing than you might have realised. But having acknowledged the importance of appearance and function, your attention can turn to practical matters.

15

1 Roofing and the Spare-time Builder

This manual has been compiled with strong conviction that, given guidance, many people are able to look after, renovate or even replace the roof on their home. However, before tackling the multifarious jobs discussed in later chapters, there are important prerequisites which are covered here. Five areas of attention are included – terminology, access, tools, statutory regulations and the role of amateur self-builders.

Terminology

Roofs come in many shapes, and are differentiated by a vocabulary which has developed over centuries. Technical terminology is used throughout, and if you come across a word which relates to roof styles, refer to the accompanying diagram. More specific terminology, however, is explained in those chapters which deal with particular rather than general matters.

Types of roof

The accompanying illustration (Fig 1) emphasises the diversity of roof types, although some of the styles shown are not common. Several are more popular abroad, such as the mansard roof attributed to the French architect, François Mansart. Purposely omitted are roofs used on industrial premises like 'northern lights', or roofs used on public buildings, such as low-pitched metal-covered structures disguised behind ornate parapets. This manual is concerned with ordinary houses and their ancillary buildings rather than mansions, and devotes most attention to common configurations such as mono-pitch, duo-pitch, and hipped roofs.

Areas of attention

The word 'roof' means many things to many people. In this manual a liberal interpretation is used which avoids traditional boundaries in divisions of labour. It perceives a roof as something with structure, a cover material, and weatherproofing detailing. It acknowledges that a roof on a house calls for the skills of tilers, carpenters, plumbers, bricklayers and several other tradesmen; it respects the fact that some tasks are beyond the reach of the amateur, and it states this quite bluntly. But it also contends that there is much which the self-build/repair enthusiast *can* undertake, as proven by many successful do-it-yourselfers. Accordingly, many aspects of building technique are embraced under one title; how much you feel is within your scope is a matter for you to decide.

Safe access

The biggest disincentive to the DIY roofer is access. At least one manufacturer of roofing tiles has confessed that amateur enterprise had always been discouraged; then added that company attitude was changing on account of the sheer numbers of self-build enthusiasts who have built roofs successfully. The fact that several manufacturers are now publishing booklets for beginners confirms the change, and product literature is often referred to in the text. But central to any concern about an amateur's ability to repair, renovate, or construct a roof is the matter of safe access.

Scaffold access

Safety is all important, and for large-scale renovation work it is essential to work from a scaffold. This presents you with two options – either you appoint a firm to erect the scaffold for you, or you hire scaffolding and erect it yourself. Curiously the layman tends to assume that by appointing a roofing contractor, he will set up the scaffolding as part of his commitment. It is true that some contractors work in this way, but others prefer to leave scaffold erection to a specialist – and subcon-

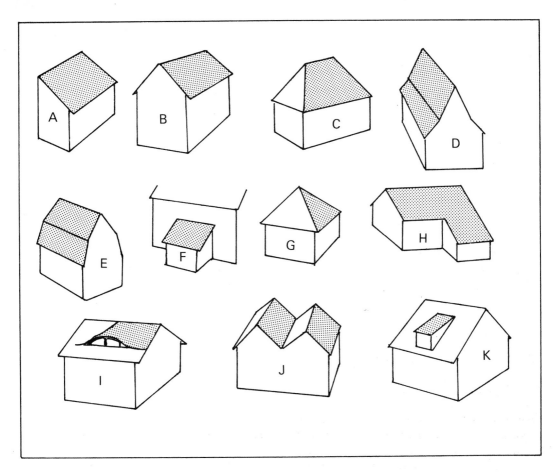

Fig 1 Types of roof and dormer window
A Mono pitch roof
B Duo pitch roof
C Hipped roof
D Roof with sprocketed eaves
E Mansard roof
F Lean-to roof
G Square four hip roof
H Duo pitch roof with dropped eaves
I Duo pitch roof with eyebrow dormer window
J Centre gutter roof
K Duo pitch roof with simple dormer window

tract the work (*see* Photo 8 overleaf).

There is no shortage of specialists listed under 'Scaffolding Erectors' in the Yellow Pages of the telephone directory. Contact can also be established by looking at scaffolds in your home area; frequently a tradeboard announces the name of the company. Don't be disconcerted if initial telephone calls find scaffolders who are more interested in servicing skyscrapers or steeples; others will be pleased to quote for small projects. Compare several companies, because in this trade you may find surprising differences between quotations.

To obtain more detailed information about scaffold hire, you can seek the advice of the National Association of Scaffolding Contractors whose address is listed in Appendix 2 (page 239). Their booklet *Scaffolders and Users Guide to Safe Access Scaffolding* is one of few sources of information. In addition to details on construction, it describes some of your obligations. For instance, as the user you must explain clearly the work for which the scaffold is required.

There is also a certification procedure; when the contractor has completed the structure, a dated certificate verifies that responsibility for the scaffold is passed over to you, together with the obligation to keep it in safe condition. Particularly important is the need to display warning notices if necessary, or remove access ladders if you leave the site unattended. Under no circumstances should you allow children to use it.

Many self-builders prefer the second strategy of hiring and erecting a scaffold themselves. If you decide to do this, safety must always be uppermost in your mind, and the access structure must meet with all

tools or paint pots cannot get kicked over the side. A tower is also useful in providing a base from which to erect a ridge-hook roof ladder.

Whether the home repairer decides to buy or hire a tower depends on several factors. Prices vary enormously, as well as the towers themselves; some are purposely intended for home maintenance whereas others are for industrial use. The 'Multiscaf' shown in Photo 9 is of interest to home repairers on account of its versatility. Though ideal for roof maintenance, it converts into trestles, a platform for wallpapering stairwells, and even a mobile workbench doubling up as the component storage rack. With many towers, storage is a problem – a point which should be checked by the prospective buyer. If the extent of your home repairs merits buying a tower, check features like adjustable feet to cope with ground variation, castors for mobility, outriggers for stability, handrails and a full complement of base and toe boards.

Photo 8 Don't hesitate to adopt the strategy followed here, in which the erection of scaffolding was entrusted to a specialist contractor

Photo 9 A safe and certain platform bestows a feeling of confidence; it also provides a base from which to set up a roof ladder

requirements. Your main source of guidance will be the booklet published by the National Association of Scaffolding Contractors. You must assume all responsibility and will be subject to statutory regulations which the booklet lists. Information is also contained in British Standard 5973, 1981: *Code of Practice for Access and Working Scaffold Structures in Steel.* Without safe access, everything which follows is folly; if in any doubt, appoint a scaffold contractor.

Mobile towers

In recent years, there has been a growing interest in sectional towers. Provided you recognise that they are not an alternative to scaffold, this form of access is ideal for simple repair jobs. As Photo 9 shows, the worker is safely enclosed by rails, and a base platform bounded by vertical 'toe boards' ensures that

Photo 10a Procedures in the erection of the 'Multi-scaf' access tower. Solid ground is essential for safety; a concrete path or a paved area is ideal

Photo 10b The base sections must be checked with a spirit level; the screw fittings on each leg facilitate the levelling procedures

Photo 10c Short planks, later to be used for the crow's nest, provide an ever-increasing work height as each section is assembled

Photo 10d Kick boards are finally positioned around the platform, while the additional outriggers shown here improve the stability. Note the tower can be fitted with locking castor wheels for better site mobility. However, with this option, the maximum recommended tower height is usually lower. Castors do not provide as much stability as platform bases which spread the load more effectively

19

For the occasional user, hiring might be a preferable strategy. The mushroom growth of tool and plant hire centres has been a welcome feature in recent years. The amateur hirer can now obtain all kinds of sophisticated equipment, but unfortunately there is not always guidance about its safe use. If you decide to hire a tower, the following points concerning its erection and use must be carefully noted.

Seeing an unattended tower blow over in a strong gust of wind is a sharp reminder of a danger to passers-by as well as to the builder; stability is all important. To begin with, the ground must be firm, and short planking is essential for distributing the load if the feet rest on soil. In no respect should the structure resemble Pisa's tourist attraction, and a spirit level is needed when base sections are laid in position. Step-by-step erection is shown in Photos 10a, 10b, 10c and 10d, and there are no short cuts. As regards height limits, you are governed by several factors, particularly the type of feet which are fitted. With a sound base, it might be useful to fit lockable castor wheels so the tower can be moved without dismantling. This is a particular help when painting a soffit, or cleaning a run of eaves gutter. In respect of the 'Multiscaf' tower, height should not exceed three times base width if castors are fitted. However, if you use base plates instead of castors, maximum working height can be increased to four times base width – 4.8m (16ft). Further stability is provided by outriggers which should be added whenever possible, particularly when working near the maximum recommended heights. For other products, safety limits vary according to design, and you should make certain that a hire company gives performance information on its access equipment. With extra wide stabilisers and special structural frameworks, some heavy-duty industrial towers give access heights up to three times a DIY model.

Avoid working in windy conditions, and in situations where greater support is needed do not hesitate to attach the tower to parts of the building. For example, 'Multiscaf' towers should be tied in at 2.4m (8ft) intervals whenever additional stability is needed. Even at low levels, this precautionary measure is worth considering. When climbing the tower without the aid of a ladder, always use an inside corner; on reaching the final platform, if you remove planks to gain final access, this contains your weight better than climbing around the outside.

Ladder access

It is assumed that you will already know rudimentary safety points about ladder design and use. When required for access to a scaffold, you should always tie a ladder sec-

Fig 2 The recommended inclination of a ladder is 75° and this can be created easily as shown here. It is also a wise measure to anchor a ladder base with stakes and rope, and to distribute its weight on a board when the ground is soft

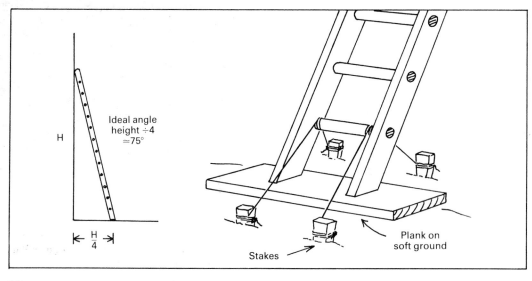

urely at the top; this is especially important when used as the means for lifting materials such as tiles on to a roof. Moreover, it should project above the top staging by at least 1.22m (4ft) to give a hold support when stepping from ladder to platform and vice versa. The last 'landing rung' should also be situated at the same level or slightly above the platform.

Access to the eaves or verges on many houses can be achieved with a ladder, although not many tasks can be performed without further equipment. Nevertheless, a ladder is fine for simple tasks like clearing an eaves gutter outlet. However, even with a 'five-minute' job, anchorage at the base is worth considering as shown in the diagram (*see* Fig 2). It is all too tempting to rest the top of a ladder against the eaves gutter, which is not good practice. Gutter joints can thus be disturbed, and it is not unusual for plastic fittings to shatter. An inexpensive answer is to fit special brackets or 'stand-off stays' to the top of the ladder as shown in Photo 11. These are available in self-assembly kits, and provide important stability.

Photo 11 Resting a ladder against eaves gutters often causes breakages; add-on wall brackets provide a suitable answer

Roof ladders

When building a new roof, or re-covering an existing one, all major works are conducted from a scaffold. At a later date, however, you may need to gain access to attend to a broken slate, or a detached ridge tile. This is usually carried out with a 'cat ladder' – sometimes referred to more simply as a roof ladder. In addition to providing access, this has the effect of distributing your weight over the roof to prevent damage. A roof ladder is pushed up the roof slope on its wheels, and then turned on its back so that the hook portion is held by the ridge. To do this, a secure stance is needed, and a tower platform is recommended. Experienced roofers fix a roof ladder in position with deft skill, and some do this from the top of an ordinary ladder. But this practice is *not* recommended; a roof ladder requires difficult manhandling which is best effected from the security of a tower or scaffold (*see* Photo 12). Once placed, total reliance on the ridge hook is unwise, and additional anchorage is recommended. A measure of initiative is needed to seek out

Photo 12 A roof ladder can be safely located from a temporary scaffolding or access tower

21

Photo 13 A bolt-on roof hook is an inexpensive way to convert an ordinary ladder for roofing repair work

Photo 14 Trestles provide an ideal platform for clearing the eaves gutter or painting soffits and fascias on a single storey building

tying in points, but do not hesitate to explore every possibility. Disregard the questioning looks of onlookers who wonder why you are not prepared to place all your trust on the ridge hook. They may not have seen a ridge tile collapse. . . .

The occasional home repairer is unlikely to purchase a roof ladder, and may prefer to hire one. Alternatively, an inexpensive kit from Burton Wire and Tube Co comprises adaptor stays which you can bolt on to the rungs of an ordinary ladder. Known as the 'Universal Roof Hook', this useful accessory is shown in Photo 13.

Trestles

In single-storey buildings such as bungalows, garages, or small extensions, access poses less of a problem. As shown in Photo 14, a pair of trestles and some scaffold planks provide an ideal platform for work at eaves level. Trestles can be home-made if your carpentry is good, but do not use a portable workbench as a trestle because it is not usually made to

Photo 15 Repairs to chimneys or stacks require a safe means of access and a firm work platform

be weight-supporting. The Hirsh 'Iron Horse' and 'Workgrabber' metal trestles shown in Photo 14 are sturdy self-assembly trestles from Sumaco, and offer a distributed load capacity of 900kg (over 2000lb). These provide support for the heaviest human, and any amount of additional loading.

Chimney staging

Perhaps the most challenging of all renovation tasks is the removal of a large chimney. Attention to brickwork around the stack, or replacement of the lead flashing at its base is also demanding in terms of access. There is only one safe answer, and this involves the construction of a chimney work platform (*see* Photo 15). On one hand, this might be erected by a specialist contractor, or alternatively a purpose-designed rig can be used. The 'Readyscaf Chimneydeck' shown in Photo 16 (manufactured by Ready Scaffold-

23

Photo 16 For safe work on chimney stacks, the 'Chimneydeck' is a lightweight structure which can be carried up a roof ladder and assembled by one person

ing Ltd) is one of the neatest answers. Its erection is remarkably straightforward, and has been described as the 'one-man portable chimney scaffold unit'. The lightweight sections can be carried up a conventional roof ladder, and assembled on roofs of any slope. A clinometer supplied with the Readyscaf is used to measure pitch angle, and the work platform can then be adjusted at ground level to match the steepness of the roof. Accessories are also provided to enable the scaffold to be used on stacks situated elsewhere on the roof slope rather than at the ridge.

Chimney scaffolds, like the Chimneydeck, are becoming increasingly popular on account of their versatility and convenience. However, if not available in your local tool-hire shop, make enquiries at plant and scaffold hire specialists.

Tools and ancillary equipment
Hoists

In installing a new roof covering, the ambitious refurbisher and self-builder will be faced with the arduous task of lifting tiles or slates on to the scaffold. In particular, concrete tiles, which account for a very large proportion of today's new roofs, are remarkably heavy. Whereas the traditional approach was to employ muscle, more and more roofing contractors are buying or hiring mechanical lifts. These take many forms, but the example shown in Photo 17 illustrates a hoist being used when reroofing a large three-storey Victorian town house. The substantial platform accommodates a large number of tiles per load, and saves sufficient manpower time to more than pay for its hire. The size of a project, the amount of time you can spare, and your physical fitness are factors which may commend machinery rather than manpower.

Chutes

In confined spaces, a reroofing project can be dangerous. Discarded tiles must not be allowed to tumble indiscriminately from roof level, and time spent to arrange their speedy but safe disposal is time saving. If you decide to hire a skip, a convenient arrangement is for a chute to transport unwanted masonry direct from scaffold platform to the refuse container. Ingenuity with scaffolding and planking is sometimes used to make a chute, but there is increasing interest in purpose-made systems. The 'Europachute' shown in Photo 18 (Europalite Ltd) is finding its way into more and more hire centres, and is becoming standard equipment with roofing contractors who work in urban areas. The chute is assembled using moulded polyethylene sections, and units are held firmly together with high-tensile steel chains. If the route to the skip or waiting lorry is indirect, gentle curves can be formed by passing a rope through the centre of the assembly and anchoring it firmly at each end.

Covers

Traditional tarpaulins made of tarred canvas have given way to PVC-coated nylon sheeting as shown in Photo 19. The Yellow Pages usually include an entry under 'Tarpaulin dealers' if your nearest hire shop is unable to help. Any projected opening up of a large roof requires the use of a suitable temporary

Photo 17 (Top) The self-builder would consider the hire of a mechanical hoist as an effortless way of transporting slates or tiles to roof level

Photo 18 Available from major plant hire companies, purpose-made chutes are essential when re-roofing work is undertaken close to roads and walkways

Photo 19 A tarpaulin cover can be hired cheaply for large scale repairs and replacement; note the safe stacking of tiles against a mesh restraining fence and kickboard

Photo 20 Specialist tools for roofing – ripper, zax, bench iron, slater's hammer, break iron and tile cropper. Other tools include knee pads and slate cutter

covering, and thin-gauge polythene sheeting is rarely suitable for anything on a large scale. A purpose-made tarpaulin includes eyelets around the edges, and adequate tying is essential, especially on an exposed site.

Tools

As with any trade, there are a number of specialist tools for the tradesman, and Photo 20 shows some examples. These are available from specialist tool shops, and retailers like Mustang Tools and Dimos Marketing (UK) have a nationwide reputation for roofing equipment. Without questioning the value of these items, it might be added that a multitude of tasks can be performed without them. Unlike the tradesman, the self-builder slating a 'one-off' roof is not going to suffer from working without knee pads. A leather nail pouch is indispensable to the tradesman to speed up work rate, but an amateur who is not subject to contract completion dates can manage without. No less relevant is the fact that the beginner builder will not have the skill to wield some of the professional's tools like the slater's hammer. The DIY approach to forming holes will adopt the slow but sure strategy of using a power drill or a 150mm (6in) nail and a hammer.

Cutting roofing felt, measuring and marking, sawing battens and pointing up ridge tiles are tasks calling for tools which you probably own already. However, if you need to remove the remnants of a damaged slate on an existing roof you will need a specialist tool called the slater's rip. Similarly, to cut concrete tiles you will need to buy or hire a power-driven abrasive disc machine. But these are exceptions to the general rule.

With reference to roof structure, the number of tools required depends on whether prefabricated trusses are used or whether the roof is built with a traditional purlin and rafter system. For setting out, many tradesmen use a metal roofing square which performs a number of setting out and checking tasks. However, if a trussed system is used on a simple duo-pitch roof, the total complement of tools needed is surprisingly small.

If you want to choose a new roof-covering material, you might need a clinometer. This

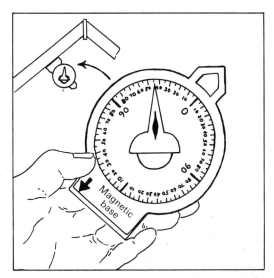

Fig 3 An inexpensive clinometer is available from specialist roofing tool stockists which measures pitch angle. Readings can be taken from either the top-side or under-side of the rafters. A magnetic base is provided for dealing with steel roof structures

is used to measure pitch angle, and an inexpensive example available from specialist tool shops is shown in Fig 3. However, there are alternatives, and calculations can be made using trigonometry. Another aid is a boxwood rule/protractor made by Rabone Chesterman; an integral spirit level and graduated brass knuckle joint indicate pitch

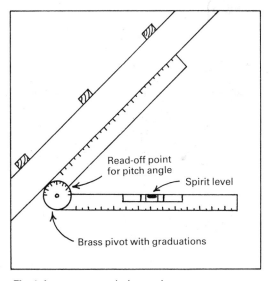

Fig 4 A protractor rule is another way to measure roof pitch. One arm is placed against the underside of a rafter; the other is set in a horizontal plane using its integral spirit level. Graduations on the brass pivot are marked at 5° intervals

angle when the rule is held against a rafter, as shown in Fig 4.

Understandably, many tools will simplify and speed up work; but much can be done without them.

Regulations and codes of practice

A successful roof should not only fulfil functional requirements, but should also enhance the property of which it is part. This duality is reflected in the two separate approvals which are needed prior to building a new roof or commencing major alterations. On one hand, the proposals have to gain planning approval; on the other, they must comply with the requirements of the Building Regulations. If your work involves straightforward repairs or maintenance, neither is required.

Planning approval

The Town and Country Planning Act (1971) is concerned with the control of development. Under the definition of 'development', local authorities who administer the Act appear to have little control over roof coverings in respect of their visual amenity. The exceptions are roofs within a conservation area, or the roofs of listed properties which are deemed to have architectural or historical interest. Nevertheless, in so far as certain types of roof work materially affect the appearance of a building, some authorities are inclined to challenge any radical change of cover material which is aesthetically inappropiate to the location. It is arguable whether they are in a position to take action, although common sense suggests that radical changes constitute a lack of sensitivity on the part of the builder. Acknowledging that planning laws are for the benefit of the community as a whole, it is recommended that you seek advice from your local planning office in respect of major roof projects. Moreover, there is a splendid free booklet, *Planning Permission – A Guide for Householders* (HMSO), which explains procedures, obligations and exemptions.

Building regulations

The concern of building regulations is the structural integrity of buildings – for the health and safety of both present and future occupants. Many of the functions detailed in the Introduction are governed by building regulations, though it is principally in new building work where implementation of the regulations is mandatory. However, if you rebuild part of an existing roof structure, you must apply for approval at your local Building Control office. This procedure is formalised, and a Building Inspector will make periodic checks to verify that you are completing the work correctly. Although the local authority administers the regulations, the Building Control Department is wholly separate from the Planning Office; the Department may even occupy buildings in different parts of the town.

It is confusing for the self-builder and home improver to learn that there are separate building regulations for Inner London. These are set out in the London Building Acts and London Building (Constructional) By-laws. 'The Building Regulations' which apply to the remainder of England and Wales are the same in spirit but not in documentation. Moreover, 'The Building Regulations (Northern Ireland) 1977', and 'The Building Standards (Scotland) Regulations 1981' are also broadly similar to the statutory instruments for England and Wales. Moves are currently being made by central government to embrace the regulations for Inner London with the regulations for England and Wales. If implemented, this policy would go a little way towards easing a complex situation in which four sets of regulations are in force for different parts of the United Kingdom.

In order to support advice given to the reader, reference is often made to Building Regulations. However, at the time of writing, the Building Regulations 1976 (England and Wales) and amendments of 1978 and 1981 are on the point of being superceded by new statutory instruments. Whereas no major changes are anticipated in respect of roofing, the new regulations will be less precise and the functional requirements will not be specified in the detail previously stated in print. With new regulations pending but not approved by Parliament, this manual uses text reference to the 1976 provisions and their amendments.

British Standards

Most products and procedures associated with roofing are carefully documented in the publications of the British Standards Institution (BSI). Codes of Practice are drawn up after lengthy deliberation by committees of experts, and a typical method of title reference is:

BS 6367 : 1983 : *British Standard Code of Practice for Drainage of Roofs and Paved Areas*

Publications from the BSI can be purchased direct, but the home repairer would be most likely to obtain copies from a public library. Alternatively there are libraries in colleges at which members of the public can obtain permission to use certain material, such as BSI documents, for reference. If you require specific information on a topic, text references are given to the relevant BSI literature.

As a caveat, it is appropriate to mention that British Standards are continually being updated. For instance a revision in relation to prefabricated roof trusses made it necessary to make last-minute alterations to Chapter 4. (BS 5268 : Part 3 : 1985). At the time of publication, every attempt had been made to ensure that this manual was correct with regard to British Standards. However, to ensure that you are following the latest practices, it is advisable to consult the manufacturers of the roofing materials you intend to use. Of necessity, their instructional literature is promptly updated whenever there are important revisions made by the British Standard Institution. Moreover, this documentation is usually supported by reference to the relevant British Standards applicable to their product.

The role of the amateur builder

For the most part, this book is presented to the amateur builder who either sees 'do-it-yourself' as a form of creative relaxation, or perceives it as a useful way to reduce the cost of living. The question remains – how much do you do yourself? The answer is dependent on your particular range of talents, your time, your budget and your commitment. It is wise to recognise shortcomings, and to enlist the help of experts where appropriate. Never forget that in the context of building, 'do it yourself', does not mean 'try and do it *all* yourself'. Even your local professional builder assigns tasks to skilled tradesmen wherever appropriate.

Some of the chapters which follow describe tasks which you might decide not to follow. Moreover, in some parts of the text, you are wholly discouraged from attempting self-help. But 'knowing' rather than 'doing' enables you to have an insight into roofing, and an opportunity to monitor closely the work of tradesmen whom you might appoint.

2 Recreational and Storage Buildings

For many householders, introductory lessons in roof repair and replacement are learnt from garden sheds, storage buildings and semi-portable workshops. These structures are a good 'proving ground' on which to practise and perfect DIY skills; an error when reroofing a coal store is rather less serious than an error when attending to the roof of one's home. Also relevant is the fact that non-habitable buildings often fall beyond the legislation set out in Building Regulations. However, the criteria which deem a building exempt can confuse the amateur, and it is always prudent to consult your local authority building control office before embarking on construction or alteration. You will find that many buildings are classified as 'partially exempt'.

'Partial exemption' concerns single-storey buildings used exclusively for recreational or storage purposes, and Schedule 2 of The Building Regulations 1976 makes reference to summer houses, poultry houses, aviaries, greenhouses, conservatories, orchard houses, boat houses, coal houses, garden tool sheds, potting sheds or cycle sheds. It is unlikely that these buildings are of interest to the local authority if they are wholly detached, not less than 2m (6½ft) from any building within the same boundary, and with a floor area not exceeding 30 sq m (232 sq ft). As a consequence, roofing materials which do not satisfy Building Regulations with regard to habitable dwellings can be used on recreational or storage buildings. Hence you are at liberty to use plastic corrugated sheeting, or cold-bonded bitumen felt. This is fortunate because coverings such as tiles or slates are usually unsuitable – the roof pitch of many garden sheds is too low. You are therefore left with a limited choice of materials, and it is important to be aware of the options, and their respective merits.

Before considering different covering materials, it must be appreciated that choice is also governed by roof structure. If it is your intention to use rigid sheeting such as corrugated plastic, the roof will need no more than a support framework. On the other hand, if you wish to use a flexible material such as bitumen roofing felt, a solid base is needed such as tongued and grooved boarding or WBP plywood (*see* page 55). Where portability is required, for example a sectional shed, a boarded roof covered in bitumen felt is more easily dismantled and re-erected than a corrugated roof fixed to a framework. In the following list of typical roof coverings, a division based on the supporting structure is used.

Rigid sheet materials

According to British Standard 5472, profiled sheeting embraces all products formed longitudinally with regular spaced shaping to give troughed or corrugated cross sections. With the advent of the industrial estate, we are well aware that material falling under this designation has achieved a notable prominence as roofing or external-wall cladding on factory units. At the same time, a simpler type of corrugated sheeting has long been in use for the allotment hut, the garden workshed, the carport or the chicken house. Whereas corrugated sheet materials are used for domestic housing in Alpine regions, especially in France, they are rarely used in this country. Photo 58 (page 108) however, might be seen as a pointer to changes in the future.

Although profiled sheeting is made in colour-coated aluminium or purpose-pressed steel, this is used principally for industrial buildings. Sheeting used for small buildings is usually made from fibre cement, zinc-coated steel, plastic, or bituminous fibre. British Standard specifications govern the shape and dimensions of profiles, thereby giving standardisation across product ranges. Thus it should be possible to construct a roof in solid corrugated sheeting, and introduce a matching plastic panel at a later date. This is particularly useful if a natural light source is sub-sequently needed in a solid roof.

Fibre cement

Manufacturers who produce fibrous synthetic slates are likely to make corrugated sheeting as well. For example, Eternit TAC and Tunnel manufacture both slates as well as sheeting. Originally the product was made from Portland cement with a reinforcing layer of asbestos – a sheeting which is still available, in spite of general concern about asbestos products and health. This is commonly used on agricultural stores and outbuildings. It is tough, though brittle, and to an extent has been replaced by an asbestos-free version in which natural and synthetic fibres provide the reinforcing bond for the Portland cement. The material is easy to paint using a conventional exterior gloss, but pre-coloured versions can be obtained. A wide range of bolts, fixings and fittings are available, together with accessories like ridge cappings and eaves closures which keep out birds and vermin. You are more likely to find these products at a builders' merchant or agricultural supplier rather than the High Street DIY shop. This roofing material is easy to install, covers a given area quickly and is one of the few coverings which can be used on low pitches. But it is not attractive, and whereas it is suitable on a Dutch barn or factory block, careful thought should be given to its suitability in residential environments.

Zinc-coated steel

Similar thoughts apply to zinc-coated steel, which is more affectionately called 'corrugated iron'. This shares the functional advantages (and ugliness) of its fibre equivalent, but can easily and effectively provide a roof on a poultry box or portable building. Its advantage is strength, and whereas it might dent it doesn't shatter like asbestos cement sheeting. Even when denied a periodic coating of paint, its life can exceed twenty-five years. Admittedly rust will eventually break through the zinc-coated surface – especially at coastal locations where salt-laden breezes hasten damage to ferrous metals. But whereas zinc-coated steel roofs last well, aesthetically they are not attractive. Furthermore, their performance in respect of thermal insulation – the ability to retain warmth in winter and to resist solar gain in summer – is negligible. As a roofing material, corrugated-steel sheeting is better suited for the homes of poultry rather than people.

With growing interest in alternative materials like aluminium sheeting, it is becoming less easy to find suppliers of this once popular product. However, TAC is one company manufacturing a hot-dipped galvanised profiled steel sheeting, and agricultural specialists or builders' merchants are the most likely stockists.

Rigid PVC

Corrugated PVC plastic sheeting is much easier to purchase, and 'never to be repeated' price-cutting advertisements often appear in weekend newspapers. Similarly the High Street DIY shop is also likely to hold stocks of plastic sheeting. Undoubtedly it is an extremely cheap roof-covering material, and together with ease of handling and simple fixing procedure, this often epitomises the first cautious steps into DIY roofing. Regrettably, some amateur builders graduate no further, which is a pity, and all too often this material is taken beyond its most suited applications. For obvious reasons it is ideal for self-build greenhouses or cloches, but when used indiscriminately, a plethora of plastic does not enhance a property. If used for a porch or carport, it is incumbent on the householders to perform the social duty of hiding its unattractive eaves finish and sharp edges with a discreet 'cover-up'. By constructing a deep fascia board around its perimeter, the overall visual effect can be tidied considerably. In this way the benefit of a light-diffusing roofing is enjoyed without offence.

The special feature of PVC roofing is its easy installation. The material can be cut to size with nothing more than a fine-tooth handsaw, and an ordinary hand drill is needed to form the holes for fixings. Its light weight makes for convenience in handling, the chance for self-collection on the roof rack of the family car and the need for a modest

Photo 21 Profiled plastic sheeting is an inexpensive covering material for conservatories, car ports, and garden sheds. It is light and very easy to install

support structure. At the same time, it must be acknowledged that the material will flex under load, and an adequate support is required to cope with heavy snowfalls. In the event, this is far less than the structure required for still heavier roof coverings. Particularly commendable is its suitability on extremely shallow pitches. With appropriate lapping at joints, some PVC sheeting can be used on roofs whose slope is as low as 5 degrees, although 10 degrees must be regarded as the minimum if you want to exploit the self-cleansing effect of rainfall.

A typical DIY product such as Novolux PVC sheeting has a 'self-extinguishing' fire rating, and in the Building Regulations (1976) the product is classified as a Type 3 plastic (Section E1). Like materials of its type, Novolux is a fragile glazing material and must *not* be walked on directly. When used solely as a roofing material, for example for a conservatory, crawling boards or roof ladders must be used to protect the PVC sheets whenever access is required. But whereas it lacks the strength to hold undistributed weight, Novolux is guaranteed for ten years against weathering.

Glass-reinforced plastic (GRP) sheeting

Two companies are involved in the production of this alternative light-diffusing plastic roofing sheet, and the profile dimensions of the corrugations match standard sheets made from steel, aluminium or asbestos cement. Different grades of GRP sheet are available, and sturdier versions are often used for roof lights on factories. Glass-fibre laminations reinforce the polyester plastic, and give the sheeting its tensile strength. Though transluscent rather than clear, GRP sheeting may be used for domestic roofing on the structures already described, but will be more expensive.

Bitumen-impregnated corrugated sheeting

Dissimilar in every respect except shape is bitumen-impregnated corrugated sheeting.

Using a heat process, this material is formed from a mixture of fibres which are saturated with bitumen. The resulting corrugated sheets are exceptionally tough, and have been known to spring back into shape when accidentally driven over by a site vehicle. This flexibility of strength allows them to be nailed without pre-drilling, and a fifteen-year guarantee is further evidence of durability. On sale in Britain under the name Onduline, the product is distributed in over sixty countries, and can endure all kinds of climatic conditions. Although not well known in the United Kingdom, it has been used for over forty years; in many countries it has been used as a roof material for housing – in addition to its more common application for warehouses, agricultural buildings or garden sheds.

This type of sheeting is available in five colours, together with an aluminium silver, and a variety of supplementary accessories are available. For example, a flexible ridge unit provides the capping to suit roofs of various pitches. A roof ventilator mounted on a corrugated section, and skylight opening windows are further useful accessories which are not found with other types of corrugated sheeting. Several specialist versions are produced to deal with specific problems. One type of sheet, for example, is particularly fire resistant, whereas another is designed especially to give maximum colour permanence.

Like the other corrugated materials described, installation is easy. All that is needed is a saw for cutting the panels, and a hammer for fixing them. Onduline sheeting is one of the most rugged materials of its type, and is ideal for lean-to stores or outbuildings. This is particularly true where the roof has a low pitch – even as low as 5 degrees. Tiles and slates may look more attractive but they *cannot* give a weatherproofing cover on pitches lower than 15 degrees. Indeed most tiles require much steeper slopes.

Using rigid sheeting

When purchasing profiled sheeting, you should enquire if any literature is available relating to its application and installation. Alternatively you might write direct to the manufacturer. Many firms produce leaflets which show how to use their product to best advantage. For example, if you wish to install Novolux rigid PVC sheet, the manufacturer, Weston Hyde Products Ltd, not only produces a booklet giving installation procedure, but drawings giving designs for projects such as a porch, carport and a greenhouse. (See Appendix 2, list of addresses.) An installation leaflet from OFIC (GB) Ltd also

Photo 22 Corrugated bitumen-fibre roofing is manufactured with colour impregnated into the material. This is tough, flexible and lightweight – and particularly suitable for low pitched roofs

gives full fixing methods and procedures for Onduline corrugated asphalt roofing. When selecting products, you should be aware that many manufacturers also produce heavy-duty versions of their sheeting. For example, Heavy Duty Novolux is preferred for a double-width carport, and although the sheeting is more costly, saving is made on the purlins which can be spaced up to 915mm (3ft) apart.

Support frameworks

With all profiled sheeting, you must ensure that there is a substantial supporting framework. With PVC sheeting, this is particularly important. Timber cross-members called purlins should be installed at right angles to the fall (ie slope) of the roof, and these provide support as well as fixing points (*see* Fig 5). These may in turn rest on larger

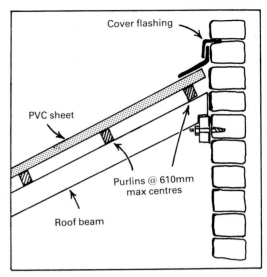

Fig 6 Structural framework for a PVC covered lean-to roof. Roof beams cut with a birdsmouth joint are supported by a timber plate which has been fixed to the wall with an expanding masonry bolt. Purlin sizes are shown in the table

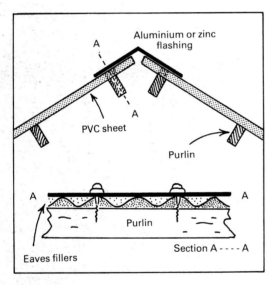

Fig 5 Timber framework on a duo pitch roof to provide support for PVC corrugated sheet. A double layer of eaves filler, placed above and below the corrugations, gives a continuous surface at the ridge to accept a cover flashing of aluminium or zinc sheeting

roof beams or rafters, as shown in Fig 6. The size of a purlin is determined by the span over which it forms a bridge, but spacing between purlins should not exceed 610mm (2ft) centres when using standard Novolux sheeting. ('Centre measurement' is the distance from the centre point of one purlin to the centre point of the next.) To find out the re-

quired size of a purlin, Novolux gives a table in their installation leaflet as follows:

Purlin sizes (For supporting Novolux rigid PVC sheet)	Maximum span of purlins (between roof beams)
38 × 75mm (1½ × 3in)	1.65m (5ft 5in)
38 × 100mm (1½ × 4in)	2.19m (7ft 2in)
38 × 125mm (1½ × 5in)	2.73m (8ft 11in)
38 × 150mm (1½ × 5⅞in)	3.27m (10ft 8in)
38 × 175mm (1½ × 6⅞in)	3.81m (12ft 5in)
38 × 200mm (1½ × 7⅞in)	4.34m (14ft 2in)
38 × 225mm (1½ × 8⅞in)	4.87m (16ft 0in)

(Based on information from 'Novolux, for improvement to your home and garden', and reproduced with the permission of Weston Hyde Products Ltd.)

If you are roofing a small shed, you may decide to use purlins alone to span from end wall to end wall, although the addition of an intermediate rafter may lower your final costs by reducing the size of purlins required for the divided span.

When using a sturdier material such as Onduline, the recommended support frameworks vary according to roof pitch. If the pitch angle is 15 degrees or more, purlins at 620mm (2ft) centres are required. Between 10 and 15 degrees, a 450mm (1ft 6in) spacing is needed, whereas a solid roof deck is needed if Onduline is used on pitches between 5 and 10 degrees.

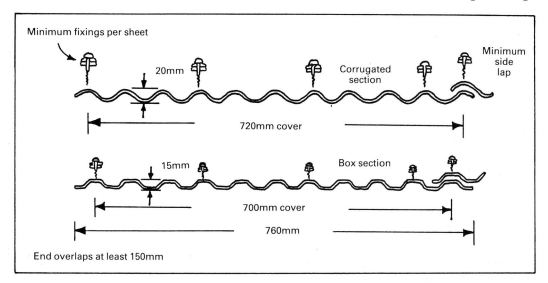

Minimum fixings per sheet

20mm

Corrugated section

Minimum side lap

720mm cover

15mm

Box section

700mm cover

760mm

End overlaps at least 150mm

Positioning and cutting

When you work out the position of sheets, ensure that they are the correct way up as shown in Fig 7, and that overlaps at the ends are no less than 150mm (6in). Side overlaps of a single corrugation are adequate for all but the most exposed sites. As shown in Photo 23, PVC panels can be easily cut using a fine-toothed saw, such as a tenon saw held at a low angle. Giving full support while the sheet is cut is important – especially at the cutting edge. An ordinary wood saw is also recommended to cut Onduline sheets, whereas metal or glass-reinforced fibre profiled sheeting requires a hacksaw.

Drilling and fixing

All forms of sheet roofing must be adequately fixed to prevent wind uplift. Similarly, in roofing, the fixing must always be made on the peak of a corrugation, and at the points where adjoining sheets overlap. Thereafter, frequency of fixings varies according to the product; with Novolux sheeting, there should be at least four points of fastening across the width in addition to the overlap fixing. With regard to Onduline and zinc-galvanized sheet, it is recommended to fix *every* corrugation at the ends of sheets, and *every other* corrugation elsewhere.

PVC sheeting is attached using special screws with waterproofing washers and caps (*see* Photo 24). Pre-drilling is necessary, but

Fig 7 Novolux box and corrugated profile PVC sheeting – lap detailing, minimum fixing requirements, and effective coverage widths

Photo 23 Given the appropriate support, plastic roofing materials can be cut to size with a tenon saw

35

as shown in Photo 25, this is easily accomplished with a hand drill and a not too sharp drill bit. A power drill can also be used, and a masonry bit will also form the hole satisfactorily. However, the point to stress is that PVC sheeting must be held with *screws,* and you should make sure not to overtighten them. In contrast, Onduline, like metal sheeting, is fixed with nails. Purpose-made large-head galvanised nails are used on zinc-coated sheets, whereas Onduline manufacture a PVC-headed nail for bitumen sheeting. On agricultural buildings, purlins are sometimes made in steel, in which case you would attach the sheeting with a galvanised hook bolt.

Ridges, abutments and eaves

On a duo-pitch roof, the ridge must be waterproofed with some form of capping. If you look at the finish on farm buildings, you will note that most fibre-cement systems include ridge cap units. The Onduline range also includes ridge pieces as accessories in colours to match the sheeting. If you use PVC sheet, aluminium-flashing material is needed at the ridge; alternatively you can use 'Metiflash' zinc/lead alloy flashing manufactured by Metra Non-Ferrous Metals Ltd.

On a lean-to building, a flashing strip and cover flashing piece are needed, and different products are described in Chapter 7. Of particular interest to users of PVC sheet is the corrugated PVC flashing strip made by Cavity Trays Ltd (*see* Photo 112, on page 186). This clear plastic covering, described on pages 186-7, is made in corrugations to suit standard profiles, and gives a much neater finish than inaccurately applied flexible sealing strips.

At the eaves, the profile of corrugated sheeting provides an entry point for birds, vermin and insects. Most manufacturers include filler units as an optional fitment, and you would be ill-advised to omit this.

Photo 24 (Top) PVC sheeting should be fixed with screws. Purpose-made caps and washers complete the attachment with a weatherproof seal

Photo 25 A hand drill is all you need to form the fixing holes on profiled plastic sheeting. But only drill on the *peaks* of the corrugations

Non-rigid roof covering

Bitumen felt

A point repeated in several chapters is the fact that there are many *different* types of roofing felt. For example, hessian-reinforced felt which is used underneath tiles on modern pitched roofs is wholly different from felt which has been manufactured with a surface layer of mineral chippings. When covering the roof of a shed, it is most important to use the appropriate material. You should also be aware of the different methods of bonding each layer of felt. In Chapter 3 on flat roofs, reference is made to hot bonding which is necessarily used on buildings governed by British Standard provision, and conditions laid down in Building Regulations. If a building requires a long-lasting cover comprising several layers of felt bonded by molten bitumen, you should appoint a roofing specialist. However, for most recreational and storage buildings, it is acceptable to use a cold-bonding process; this is ideal for DIY projects, and all you need is a wide-bladed paint scraper or soft broom as an applicator.

Self-adhesive bituminous shingles

For a more attractive roof covering, self-adhesive bituminous shingles are ideal for buildings such as summer houses, chalets and boat houses. Several products are available, and examples like Bardoline shingles (OFIC [GB] Ltd) are made with a glass-fibre base to give strength to their impregnated bitumen coatings. Cover granules give an attractive texture, and several colours are available.

The chief disadvantage of bituminous shingles is their unsuitability on low-pitched roofs. Some products are considered suitable on pitches down to 10 degrees, but others require roof slopes in excess of 20 degrees. Notwithstanding their pleasant appearance and suitability for DIY installation, you should check specifications in the manufacturer's literature when seeking a cover material for a low-pitched roof.

Photo 26 Glass-fibre based bitumen shingles are self-coloured and ideal for low pitched roofs down to a 10 degree slope. Installation is not difficult, and some types feature self-adhesive backings for effective bonding

Installing cold-bonded felt roofs

Again it is pleasing to note the awareness of manufacturers to the needs of amateur home improvers. Acknowledging that many house-holders only need a small quantity of roofing felt to re-cover a tool shed, hen house or rabbit hutch, Ruberoid market a DIY roofing kit which comprises 5m (16ft 5in) of felt, a cutting knife, clout nails and detailed instructions. Permanite are also eager to assist the amateur, and their excellent guide, *Fixing Bitumen Roofing – Using Aquatex Roofing Felt Adhesive* is well known. Copies of this booklet are available on request from Permanite, whose address is listed in Appendix 2. A helpful feature in this guide is the tables which prescribe the correct type of felts and fixing for roofs of different pitches, and different types of support decking.

Decking

Continuous support – referred to as a 'deck' – must be given to both felt and bitumen-shingle roofs. Tongued and grooved boarding was the traditional choice, but in modern practice, timber sheeting is preferred. This is more stable, easier to install, and likely to be less expensive. Chipboard (sometimes called particle board) is cheaper than WBP plywood and has become increasingly popular. But you *must* use high-density (roofing grade) chipboard; this is available with tongue and groove edging so that adjacent panels interlock to give a sound decking. On a large roof, the decking will need to be supported on rafters, and this topic is discussed in the following chapter. But for a small building such as a 2.4 × 1.83m (8 × 6ft) shed, one cross purlin centrally placed would be sufficient to support 18mm (¾in) high-density chipboard. You should add softwood edging battens around the perimeter of the decking to provide a good fixing point for the 'drip' edging. Finally, you should note that the roofing felt must be laid on a deck which is *completely* dry.

Felt layers

You should install three layers of felt on a roof which is expected to give long service; two layers of felt is only recommended as a short-term provision. In both cases, the top layer and drip surround require the type of felt which is covered with mineral granules; under-layers should be a plain-surfaced felt which contains a fibre or polyester base. If the pitch of the roof exceeds 10 degrees, it is usually recommended to lay the felt longitudinally in line with the slope. The edges of the felt lengths should be nailed and overlapped at the sides by at least 50mm (2in). Joins in layers must be staggered so that the edges never coincide. On roofs flatter than 10 degrees, the run of the felt should be *across* the roof, and you will commence by laying the first layer along the eaves. Again, the joins must be staggered.

Installation (See Figs 8A-L)

On a chipboard or ply deck, all joints should be covered with 150mm (6in) strips of felt held in place with cold-roofing compound.

Fig 8A–8L DIY bitumen roofing using cold compound felt adhesive

A three layer system is best on most built-up felt roofs, but on a temporary building such as a small shed, a two layer felt roof is acceptable. The under-layer should be either a smooth faced *fibre based felt*, or a *polyester based felt* which is more tear resistant but higher priced

The top layer, perimeter upstands and drip finishes must be made from a *mineralised surface felt*. The procedure illustrated here assumes that the shed has a deck of either 19mm (¾in) WBP plywood or roofing grade chipboard. A perimeter finish is also necessary, and softwood battens provide this edge reinforcement

A Joins in the decking material should be covered with 150mm (6in) strips of fibre based felt, held in place with a proprietary cold compound adhesive such as Aquatex by Permanite. On a roof whose pitch is less than 10°, the first layer of underlay felt is laid *across* the roof, starting at the eaves
B Minimum horizontal and vertical laps are shown in the drawing. The base cover of fibre felt, referred to as the 'underlay', should be spot and strip bonded to give a 'partial bond' on decks built from plywood or chipboard. Stagger nailing at 150mm (6in) centres is preferred on a timber boarded roof. Fixing along the upper edge, however, is achieved with galvanised clout nails, 19mm (¾in) long and with extra large heads
C Edges are waterproofed with strips cut from mineralised felt which is folded to form 'drips'. Drips are cut from a roll in widths of either 200 or 250mm (8 or 10in) according to the cover required
D Drips are lightly scored and folded to form a 50mm (2in) double thickness edge
E To waterproof corners of the roof or to form waterproof joins on a horizontal run, the drips should be cut to give projecting lugs. To achieve good bonding when using the adhesive, mineral granules should be scraped from the joining surfaces
F The first stage in fabricating a drip is to fix all drip strips to the softwood batten around the roof edges using clout head nails. The reverse side of the felt should face outermost

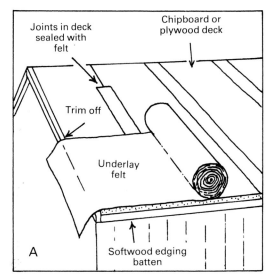

Joints in deck sealed with felt

Chipboard or plywood deck

Trim off

Underlay felt

Softwood edging batten

A

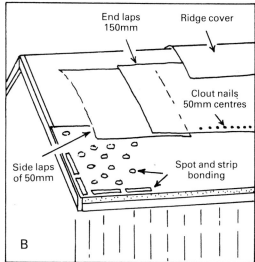

End laps 150mm

Ridge cover

Clout nails 50mm centres

Side laps of 50mm

Spot and strip bonding

B

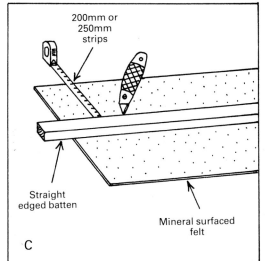

200mm or 250mm strips

Straight edged batten

Mineral surfaced felt

C

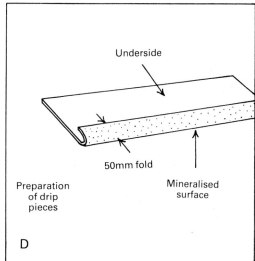

Underside

50mm fold

Preparation of drip pieces

Mineralised surface

D

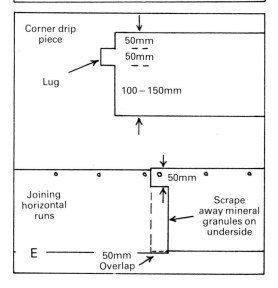

Corner drip piece

50mm
50mm

100 – 150mm

Lug

Joining horizontal runs

50mm

Scrape away mineral granules on underside

50mm Overlap

E

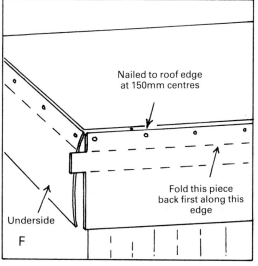

Nailed to roof edge at 150mm centres

Fold this piece back first along this edge

Underside

F

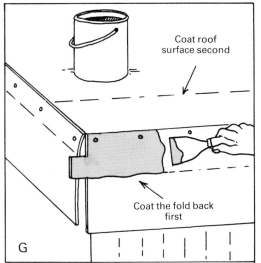

Coat roof surface second

Coat the fold back first

G

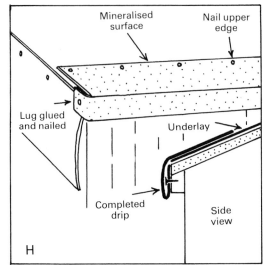

Mineralised surface

Nail upper edge

Lug glued and nailed

Underlay

Completed drip

Side view

H

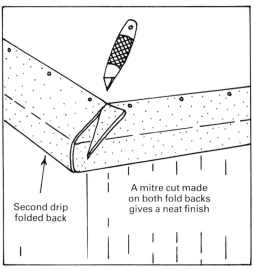

Second drip folded back

A mitre cut made on both fold backs gives a neat finish

I

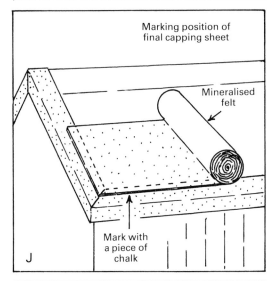

Marking position of final capping sheet

Mineralised felt

Mark with a piece of chalk

J

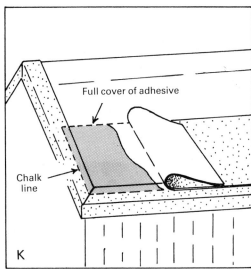

Full cover of adhesive

Chalk line

K

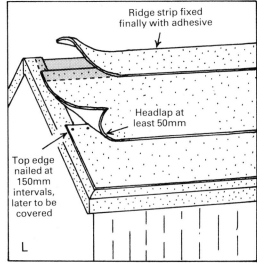

Ridge strip fixed finally with adhesive

Headlap at least 50mm

Top edge nailed at 150mm intervals, later to be covered

L

You should spread this along the joins with a decorator's knife. The first layers of felt are then fixed in place by coating the compound around the perimeter, and spots of adhesive distributed liberally over the remaining surface. On a boarded roof, the first layer should be stagger nailed at 150mm (6in) intervals instead of bonded. Strips should be cut from across the roll, 200mm (8in) wide, to form the drip skirt. These should then be scored, cut to give corner overlaps and fixed around the perimeter with large-headed galvanised clout nails. For the moment these are left to hang while the next base layer is added. A complete cover of adhesive is needed to bond the remaining layers and, depending on weather temperature and your choice of compound, this will probably take about twenty minutes to get tacky.

Before you fix the mineralised top layer, each drip section should be coated with adhesive, and folded back on itself as shown in Fig 8H, to produce the perimeter 'drip' skirt. The fold-backs will then be topped with the last layer of felt. This should be cut exactly, positioned temporarily, and its cover area marked out with chalk to show the area which needs the final coating of adhesive.

On lean-to roofs, the upstand can either be formed with mineralised felt to give a weather-proofing cover, or finished with a non-ferrous flashing. On a duo-pitch roof, the ridge can be covered with additional finishing strips of mineralised felt.

Installing bitumen shingles

For full fixing details, you should consult the technical literature from the manufacturer; different products involve different fixing recommendations, and this summary is only intended to give a general picture of the work involved.

Full support is again needed from a rigid decking, which should be constructed according to manufacturer specifications. For example, chipboard is not suitable for Ruberglas 105 slates, and exterior-grade plywood or boarding must be used instead. When complete, the decking is used as a large drawing board, on which to chalk the setting-out marks; these ensure that shingles are laid in horizontal alignment. Guide marks are also needed from eaves to ridge so that the vertical joints ('perps') are lined up.

Bitumen shingles are made in strips of three or four with a backing of special adhesive. Manufacturers specify nailing at intervals, and the units bond to the roof automatically by the warmth of the sun. However, in cold weather a heat torch may be needed to activate the adhesive. With Ruberglas 105 slates, the manufacturer also recommends a smear of mastic over nail heads, and the addition of small patches at the corners of each tile to resist wind lift. If you need to form valleys where roofs intersect, or enclosures at the verge, mineralised felt can be cut to shape and attached with clout nails and cold-compound mastic. Flashing strips can also be made up from mineralised felt to provide effective weatherproofing if a roof has chimneys, abutments and vent pipes. These features would not normally appear on the type of buildings discussed in this chapter, but it musn't be forgotten that bitumen shingles are also used on chalets and holiday homes. Most of the covering materials described earlier do not comply with Building Regulations, but it should be noted that glass-fibre bitumen shingles are acceptable on habitable dwellings if certain conditions are observed. Of particular importance to the amateur is the fact that their installation requires few tools, and no high degree of building skill.

G Adhesive is applied to the upper part of the drip, in readiness for folding it back on itself over the nail heads. A coating on the roof surface is added next to receive the remaining part of the edging strip
H The drip is given additional anchorage on the face of the roof with clout nails driven at 150mm (6in) centres along its edges. Lugs are folded around at corners, glued and nailed
I After an eaves fold over, drips along the verges are folded next using adhesive and nails for anchorage. At corners, overlap is avoided by cutting mitres
J When drips are complete, the final cap layer of mineralised felt is cut so that it laps at least 50mm (2in) over the drip edgings. Chalk markings are made to define the entire area which needs to be coated with adhesive
K Whereas the first underlay was spot fixed, final cap layers must be fixed using a complete cover of adhesive. To give an example of a typical covering capacity of cold compound adhesive, one litre of 'Aquatex' will cover approximately 1.5 sq metres (1.8 sq yds)
L Only the upper edges of the cap layer are clout nailed, although the heads will be hidden by subsequent layers of mineralised felt. The horizontal laps should be at least 50mm (2in), A ridge piece is added finally, and attached solely with adhesive

3 Flat Roofs – Structure and Cover Materials

It is important for the home improver or self-builder to recognise the differences between flat and pitched roofs. To begin with, their basic structure is dissimilar. The structure for a flat roof has to be built in such a way that it provides continuous support to its cover material – a structure referred to as 'decking'. On the other hand, most pitched roofs only support covering materials via battens spaced at intervals. Occasionally a pitched roof may be boarded too, but this is unusual, except in areas where the weather is severe. The contrast in structure is also accompanied by differences in their respective covering materials. For example, a pitched roof is usually covered with small overlapping units such as slates or tiles; these would not keep out the weather on a flat roof, and a continuous sheeting is needed instead. Two technologies are thus involved, and in turn the products are different. In consequence, the roofing industry is sharply divided; manufacturers whose names are well known in the context of pitched roofs are not linked with products for flat roofs – and vice versa.

The long-term performance of flat and pitched roofs is also dissimilar on account of the different speeds at which rainwater is discharged. There is less likelihood of weather damage on a roof where the discharge of rainwater is rapid, and pitched roofs have a clear advantage in this respect. However, flat roofs are cheaper to build, and are not liable to failure, provided the design, choice of material and workmanship standards are followed. The Building Research Establishment studies show that nearly sixty per cent of failures are caused at the design stage, due to incorrect design detail. Where problems occur, the penalty can be severe, and Photos 28a and 28b show houses in a new town whose flat roofs needed frequent attention. To effect a successful long-term cure, pitched roofs were constructed over the flat roofs at considerable expense. This alteration, known as 'pitching up', was deemed necessary only a handful of years after the properties had been completed.

Lastly, the differences between the two approaches to roofing have particular implications for the amateur builder. If you decide to build a flat roof on a porch, or home extension, you are *not* encouraged to 'do it all yourself'. Whereas there is nothing unusually difficult about completing the timber 'sub-structure', you are strongly advised to appoint a contractor to install a covering where hot bitumen will be used (*see* Photo 29). Where a roof can be covered with cold bitumen, this is ideal for DIY work, but this technique is not always acceptable. However, the advent of 'torch-on' techniques shown in Photo 30 may open up interesting new possibilities for amateurs.

'Flat' roofs defined

In all but a few cases, flat roofs in this country are not flat. On account of our generous rainfall, every effort is made to remove surface water quickly and efficiently. Indeed failing to achieve this objective places the roof covering at risk. Weak spots soon become leak spots, and residual water finds a quicker way to reach ground level – via the ceiling. 'Flat' is an inaccurate adjective, and a 'flat roof' is something to avoid.

In the Building Regulations of England

Photo 27 Tiles or slates cannot weatherproof a low pitched roof; this bitumen felt finish provides a smart answer, though its useful life will never attain the longevity of pitched roof coverings

Photo 28a BEFORE   Flat roofs often give trouble – as was the case in one of Co Durham's new towns. Hundreds of houses gave trouble not many years after their completion

Photo 28b AFTER   Rather than pursue a policy of perpetual repairing, a large scale operation was implemented in which pitched roofs were erected *on top of* the original flat roofed dwellings

and Wales, 1976, a 'flat roof' is defined as a structure whose pitch angle lies between 1 and 10 degrees (Schedule 6). A roof whose slope exceeds 20 degrees is deemed to be 'pitched', although if it exceeds 70 degrees, Building Regulations conceive it as an external wall rather than a roof. (It is then subject to the requirements for external walls tabled under 'Safety in Fire'.) Roofs, whose pitch is between 10 and 20 degrees, are often referred to as sloping roofs, although texts differ in their classification. However, the point of significance is that continuous sheet coverings are essential on all roofs whose pitch angle is less than 15 degrees. Furthermore, in regions of severe weather, a continuous covering is likely to be needed below 20 degrees. But it is around this pitch angle when some designs of tile can also be used successfully. For example, Redland 'Regent' interlocking tiles can be laid on roofs as low as 17½ degrees, provided the site is not exposed. Moreover, the Marley 'Wessex' concrete tile can even be used on slopes as gentle as 15 degrees.

Photo 29 The traditional 'pour and roll' technique for built-up felt roofs is not recommended as a DIY operation. This is one task best entrusted to a roofing contractor.

Photo 30 Many contractors are now using a 'torch-melt' technique instead of the traditional 'pour and roll' method of laying a built-up felt roof

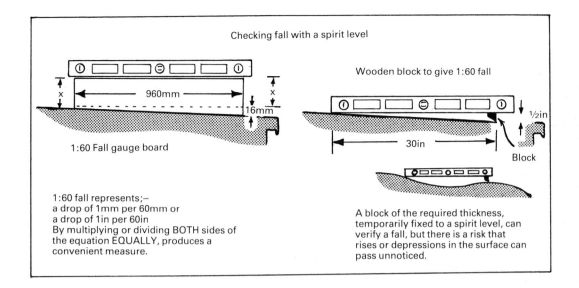

Checking fall with a spirit level

960mm

16mm

1:60 Fall gauge board

1:60 fall represents;–
a drop of 1mm per 60mm or
a drop of 1in per 60in
By multiplying or dividing BOTH sides of
the equation EQUALLY, produces a
convenient measure.

Wooden block to give 1:60 fall

30in

½in

Block

A block of the required thickness,
temporarily fixed to a spirit level, can
verify a fall, but there is a risk that
rises or depressions in the surface can
pass unnoticed.

Pitches and falls

When designing a pitched roof, its intended slope is expressed in degrees. Moreover, when building the structure its pitch angle can be measured and checked by trigonometry or using the tools described and illustrated in Chapter 1, page 27. However, when dealing with flat roofs, defining a slope in terms of its angle to horizontal is academic rather than practical. When you construct the decking, it is much more useful to have the intended slope expressed as a 'fall'. To provide an example, BS 6229 : 1982, entitled *Flat Roofs With Continuously Supported Coverings*, states that the minimum fall for a built-up bitumen-felt-covered roof should be no flatter than 1:80 (Table 1, Page 6). This means that over a horizontal run of 80mm, the roof will fall 1mm. To express this in more practical terms, you would increase both figures in proportion to give rather larger units, such as 10mm per 800mm. If you are more familiar working in imperial units, this slope can equally well be expressed in inches. For instance a fall of ½in per 40in is a representation of a 1:80 slope using convenient units of imperial measures.

Although it has been stressed that flat roofs should have *some* fall, there are variations in the recommended minimum. For example, BS 6229 specifies that a slope should be no less than 1:80. However, it is interesting that some manufacturers of bitu-

Fig 9 By multiplying both sides of the metric example by 16, the fall is represented as a drop of 16mm per 960mm. By halving the imperial example, fall is represented as a drop of ½in every 30in. Both are convenient measures if you use a metre spirit level

men felt recommend a steeper slope, and propose a minimum fall of 1:60. (As a point of interest this represents an approximate pitch angle of 1 degree. Certainly this produces a better discharge of rainwater, less chance of 'ponding' if settlement induces a small amount of flattening, and a greater chance of self-cleansing when rainwater flows over the surface. Some data sheets recommend a fall of 1:40 to give an even better performance and this is based on the Building Research Establishment recommendations. One thing you can be sure of is that a steeper slope speeds up the discharge of rainwater, reduces the likelihood of 'ponding' from roof settlement, and ultimately confers increased life to the cover material.

Measuring and monitoring fall

Whether you lay a drain, construct a sloping patio, or build a sloping roof, a spirit level is used to measure 'fall'. Noting the data on the architect's drawings, you transfer this to a piece of timber, which is cut to produce the required slope as shown in Fig 9. The equation can be increased to a manageable measure by multiplying both figures equally. Thus a 1:60 fall is the same as a 16:960 fall. If you use a metre-long spirit level, a fall of 16mm per 960mm can be conveniently represented on a board which approximates with the length of the level. If you prefer to work in imperial measures, 1:60 represents a fall of ½in per 30in run. Another way to monitor fall during construction is to tape a 16mm block to the level, 960mm from one end. But the block may get dislodged, and this method is less satisfactory than the use of a fall board.

Coverings

Several materials are used to form a continuous covering on a flat roof, and these include non-ferrous metal sheeting such as copper, zinc and lead. Asphalt is another material often used on large industrial buildings. However, in the context of domestic buildings, bitumen felt is the most common covering, and in consequence this chapter focuses on this option. Roofing felt is purchased in rolls, a metre (39⅜in) wide, and has to be laid with an overlap. In order to produce a 'continuous covering' several layers are built up with a bonding of bitumen; in the building trade, this is referred to as 'built-up bitumen felt roofing'. In addition, it is possible to buy a torchable felt like 'Britorch' from Permanite; this has a surface coating which is heated with a gas torch during application.

It is important to recognise that there are many types of roofing felt, most of which are classified in BS 747 : 1977. However, there are also a number of modern polyester/modified bitumen products which have received full approval from the British Board of Agrément and are particularly durable. It is anticipated that when British Standards are next revised, these new long-lasting membranes will be added to the listing.

Constructing the substructure

In most respects, a roof deck is structurally the same as a timber floor constructed for an upstairs room. Where it is intended to use the roof as a thoroughfare or sun deck, it will be designed with the strength of a floor to withstand frequent 'human traffic'. Usually, however, a flat roof only needs to be strong enough for 'occasional access' such as repair

work, and a cost saving can be made by designing an access-only structure.

Whereas the construction of a roof deck is within the capacity of a careful DIY carpenter, designing a roof to cover an important room or building is *not* a matter for the unqualified amateur. If you propose to build or part-build a home extension or garage, you are strongly advised to seek the services of an architect or building surveyor. The point has already been made that a roof is more than a 'lid', and loadings from wind and snow are just two of many factors which have to be taken into account when designing the structure. But having said this, there are always projects which a home owner feels are not deserving of an architect's attention, and where a confident amateur is prepared to pay the price for error. Structures such as carports are often successfully designed and built by DIY enthusiasts. It is for this reason that the following advice is presented.

Joists

A first task is to establish the dimensions of the main-deck supports known as 'joists'. Joist size is determined by the span between their supporting structure, and this information can be obtained from tables. However, a deck requires strength either for regular access, for example a sun balcony, or occasional access, such as for repairs, and separate tables deal with the different structural requirements. Information appears in Appendix data of the Building Regulations (England and Wales) 1976. This is entitled 'Schedule 6', and within the section, Table 4 is concerned with 'access only' roofs, whereas Table 5 relates to decking whose use is intended to extend beyond 'the purposes of maintenance or repair'. Within the tables, three centre spacings are listed – 400mm, 450mm and 600mm (approx. 16in, 18in and 24in). This represents the space between each joist, and measurement is taken from the centre line of one unit to the centre line of the next.

The information in the tables can be used in several ways. For instance, if you want to build a flat roof less than 10 degrees pitch over a garage, with a span width of 2.6m (8ft 6in), and to offer only repair access, the following options in joist size and spacing are listed:

38 × 100mm (1½ × 4in) at 400mm (16in) centres (maximum span 2.5m 8ft 2in)
or
38 × 125mm (1½ × 5in) at 600mm (24in) centres (maximum span 2.73m 8ft 11in)
or
44 × 100mm (1¾ × 4in) at 450mm (18in) centres (maximum span 2.52m 8ft 3in)

These options assume that you are using no more than a maximum of twenty-five joists in total, and will be using GS or MGS timber. Stress-graded timber has to be used for structural work, which means that it has been checked for weakening factors such as knots or fissures. This grading evaluation is carried out at most large timber merchants, and GS (General Structural) designated material confirms scrutiny by a qualified expert and verifies suitability for structural building. MGS (Machine General Structural) timber will have been stress-graded by machine. (These standards are laid down in BS 4978 : 1973 : *Timber Grades for Structural Use.*) If you come across SS (Special Structural) or MSS (Machine Special Structural) on drawings, these relate to slightly higher stress-graded material, with better performance and a higher price to match.

Another way to use the table information is to find out about the performance of material in a particular size. If, for example, you are in possession of 100 × 50mm (4 × 2in) GS or MGS timber and wonder if it could be used to make an access-only decking, the tables will show its span limits viz.

At 600mm (24in) centres it could be used up to 2.39m (7ft 10in) maximum span.
At 450mm (18in) centres it could be used up to 2.62m (8ft 7in) maximum span.
At 400mm (16in) centres it could be used up to 2.72m (8ft 11in) maximum span.

Notwithstanding the acceptability of timbers in terms of span strength, the spacing chosen for joists will also determine the required thickness of deck boards. For example, if you wanted to use 18mm (¾in) exterior grade WBP plywood, which is a regular stock item, this receives sufficient support when laid on joists spaced at 400mm (16in) centres. But if

you space joists at 600mm (24in) centres, 25mm (1in) material would be required instead.

It will be apparent that a number of factors are involved, and although there are several alternatives when designing a structure, a comparison of prices introduces another variable, and shows which strategy is kindest to the pocket.

Ventilating the roof void

In connection with design, it is important to be aware of the problem of condensation build-up in the space which lies between joists, and between the ceiling and roof deck. This space is referred to as the roof 'void'. The problem was scarcely acknowledged until recently, but today it is a major issue. If condensate forms *within* structural timbers – a phenomenon known as interstitial condensation – their life is rapidly shortened. Early failure of roofs is being reported frequently, and it is important that you read Chapter 7 on ventilation and insulation. The problem has led to major re-thinking in the design of roof decks, and the Building Regulations prescribed new measures (Second Amendment) in 1981. The requirements appear in Section F5 (page 8) headed, 'Specific requirements to limit condensation risks', and apply to roof voids above insulated ceilings; roofs over open porches or unheated garages are exempt.

The problem of condensation is related to insulation measures, and, depending where an insulant is placed, the roof structure is referred to as a 'warm roof', a 'cold roof' or an inverted roof (*see* Figs 10a, b and c). This is explained more fully in Chapter 7, together with the subject of vapour barriers. At this stage, it is sufficient to state that if a vapour barrier has not been installed behind the ceiling boards, vapour can percolate into the roof void. In a 'cold roof' arrangement, where an insulant has been placed at ceiling level, this vapour will cool and release its moisture as 'condensate'. Condensation will appear on cold surfaces *and* within the interstices of timbers. One way to overcome this in cold-roof construction is to install an efficient system of cross-ventilation. This is now mandatory in accordance with Section F5 of the Building Regulations.

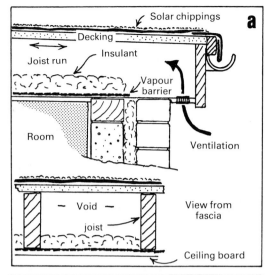

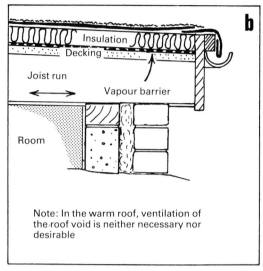

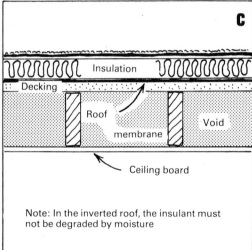

Note: In the inverted roof, the insulant must not be degraded by moisture

Fig 10a Timber flat roofs can be built either with 'cold' or 'warm' principles. The cold roof *must* have both a vapour barrier *and* an effective way of ventilating the roof void

Fig 10b The warm roof features purpose-made insulation boarding placed *above* the void; it is essential, however, to install a vapour barrier on the 'warm' side of the insulant as shown

Fig 10c The inverted roof adopts the warm roof principle, but the materials above the void are laid in reverse order. In effect, an older roof which is converted by the addition of insulation boards *on top* of the original surface follows this principle

Fig 11 Producing a cross flow ventilation in the void of a 'cold construction' lean-to flat roof where joists run at right angles to the wall of the main property. Soffit ventilation and roof exhaust vents such as 'Icovent' by Euroroof Ltd make it possible to produce a thoughput of air, but a 50mm (2in) clearance height in the void is also important

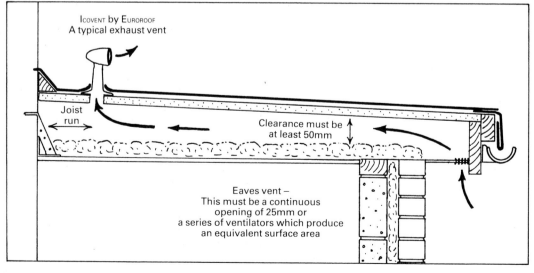

Methods of ventilating

It must be restated that a flat roof built with a vapour barrier and insulant *above* the joists does *not* require a ventilating arrangement. Ventilation is only needed in 'cold roofs', where insulating material has been placed at ceiling level, ie in the *lower* portion of the void between the rafters. If a system of ventilating is efficient, vapour will escape from the void to the outside air *before* it has time to cool and release condensate. To achieve this, openings are needed at *both* ends of the joists and it is also necessary to ensure that there is a gap of at least 50mm (2in) between the insulant and the underside of the decking (*see* Fig 11). This arrangement permits a cross-flow of air – a situation which cannot be achieved if openings are confined to one end of the void. How you make this provision depends on the run of the joists.

Joists built at right angles to a main outside wall

Many home extensions are built in this manner because it is both logical and cheaper to place joists across the narrower span. However, after decking and ceiling boards have been installed, the enclosed voids between the joists are bounded at one end by the wall of the main building, and by the eaves construction at the other. To create a throughput of air along these voids, an unbroken vent, at least 25mm (1in) wide, should be installed at the fascia extremities. Various purpose-made vents are available which incorporate a mesh to keep out birds, insects and vermin. Venting is then required at the opposite end, and on account of the wall abutment the only answer is to install mushroom-like vents on the roof. These are available from several manufacturers; for example Euroroof produce pressure-relief units under the name of 'Icovent', and these can be installed on both new and existing roofs (*see* Fig 11).

As an alternative strategy, a cross-flow can be induced at right angles to the joist run if cross-battens are nailed on top *before* the addition of decking (*see* Fig 12). These may be graduated in depth to create a fall (*see* page 50). When adopting this strategy, 25mm (1in) mesh-covered continuous openings at fascia extremities are again needed.

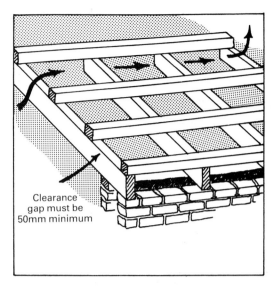

Clearance gap must be 50mm minimum

Fig 12 Cross battening is an alternative strategy for producing cross flow ventilation on a 'cold' flat roof

Joists built parallel to an abutment wall

The ventilation of the voids presents no problem if the joists are installed *parallel* to the abutment wall. Usually this means that they need to cover a greater span – and hence larger timbers will be needed. However, continuous openings can then be made in the fascias at both ends of the run to give efficient cross-ventilation.

Warm-roof alternative

Creating a system of ventilation is quite involved, and can be avoided altogether with 'warm roof' construction. As described in Chapter 7, a roof void receives warmth from a room below, provided that insulation layers are placed *above* the joists rather than directly on the ceiling boards. One way to achieve this uses Rubertherm as shown in Photo 31. Maintenance of temperatures in the voids prevents accumulated vapour cooling to 'dew point' and releasing a condensate. The warm roof alternative has become popular since the advent of sandwich-insulated deckings. On account of their contribution to structure, they receive mention here, but a fuller explanation of their role as insulators is discussed in Chapter 7. The nature and function of vapour barriers are also described later.

Creating falls

There are several ways to create a fall on a flat roof. The simplest answer is to incline the joists, but if ceiling boards are added later, this produces a sloping ceiling. In a detached garage, where joists are generally left exposed, the tilt visible on their underside is not considered unacceptable. In consequence, this strategy is often used. But where you want to avoid a sloping ceiling, joists should be placed level, and tapering lengths of timber nailed on top to produce a sloping base for the deck boards. These wedge-shaped additions are known as 'firrings', and a good timber merchant will cut these to suit your requirements (*see* Fig 13). An alternative to firrings is to buy decking which is specially manufactured in wedge-like sections. For example, 'Korktaper' from Euroroof Ltd is a cork insulant made in sloping sections to produce falls of 1:40 or 1:60; Coolag produce a similar taper system. These are used extensively in refurbishment work to give an increased fall on a failed roof structure. They can also be used on new work.

Photo 31 Installing an insulant *on top of* a flat deck is one way to avoid condensation problems. 'Rubertherm' combined insulation and roofing felt fulfils the object in old or new roofs in a single operation

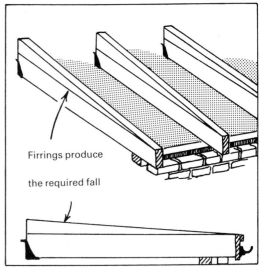

Firrings produce

the required fall

Fig 13 Lengths of timber known as firrings can be cut to the required taper at most timber yards. These are usually nailed on top of joist timbers as shown, though they can also be nailed *across* joists where the run is parallel to the wall of the main building

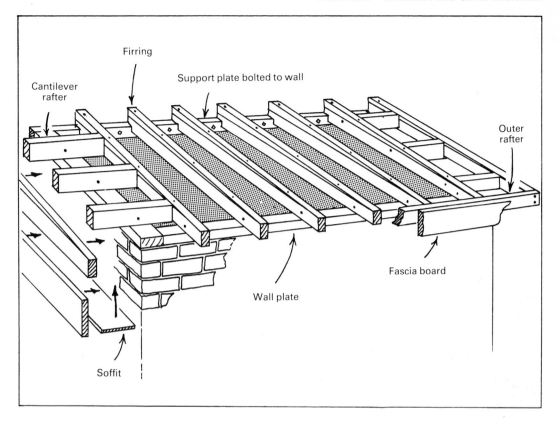

Firring

Support plate bolted to wall

Cantilever rafter

Outer rafter

Fascia board

Wall plate

Soffit

Fig 14 Structural detailing of a typical flat roof. For clarity, this is shown over a single leaf wall, which would be acceptable for an outhouse. For a habitable room, a double skin cavity wall would be built

On home extensions, the addition of firrings is the most common way to produce a fall, and the wedge-shaped pieces are usually nailed *lengthways* along the joists. However, an alternative is to fix cross-battens of diminishing depths at *right angles* to the joists. This second strategy lends itself to a system of cross-ventilation – as long as the openings formed over the joists always exceed a recommended minimum of 50mm (2in), and don't taper down to nothing. When attaching cross-battens, you must ensure that their position will create a consistent fall when deck boarding is added. This can be monitored periodically by bridging the battens with a length of timber which bears a known straight edge. This helps to avoid the creation of irregular steps which would subsequently be reflected on the boarding. By comparison, the use of purpose-cut tapered firrings is a much quicker way to create a sloping base framework for the deck material.

Installing joists

Constructing a flat-roof structure is not a difficult job for the careful woodworker. When purchasing joists, remember to specify the category of stress-graded timber; it is often best to select the material yourself and bring it away from the merchant on a sturdy roof rack or goods trailer. It is important to reject anything resembling an aeroplane propeller, and you must be ruthless and selective; twisted timber cannot be straightened. Timber is stocked in a variety of standard metric lengths, and stock positions often dictate that you accept pieces which exceed your requirement. Timber is too costly to waste, but within reason, offcuts can often be used for fascia support or strutting.

Before commencing work, you should sight down the lengths of each joist and note any slight curvature. Each length should be graded in order and laid out on the ground. When positioning joists, any rise should be on the uppermost side, and joists with the greatest curvature should be placed centrally. Fig 14 shows a typical construction, and

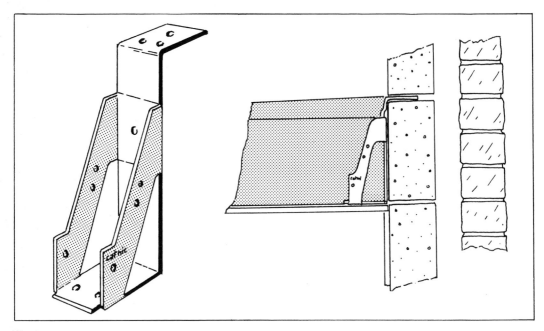

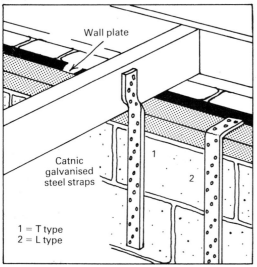

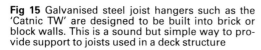

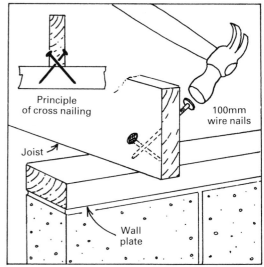

Fig 15 Galvanised steel joist hangers such as the 'Catnic TW' are designed to be built into brick or block walls. This is a sound but simple way to provide support to joists used in a deck structure

Fig 16 In modern practice, a timber wall plate must be bedded on mortar *and* mechanically fixed to the wall. 'Catnic' steel straps produce a sound means of attachment

Fig 17 Cross nailing fixes joists in place on to the timber wall plate

if you are working to drawings there will be broad similarities.

On a lean-to building, joists are usually supported on the wall of the main building with purpose-made joist hangers (*see* Fig 15). A cross-timber, fixed to the wall with expanding bolts, offers an alternative although this is sometimes left conspicuous at ceiling level. At the opposite end, the joists should rest on a 100 × 75mm (4 × 3in) timber known as a 'wall plate'. A wall plate gives a flat base, distributes roof weight over the entire wall, and provides a fixing point. A wall plate is bedded on mortar to produce a level surface, and when dry it can be anchored by driving masonry nails into its support blockwork or brickwork. However, nails can crack the mortar bed and overall, this method of fixing gives insufficient anchorage to combat wind lift. Modern practice dictates the need for a more secure system, and purpose-made galvanised straps are obtainable. For instance the products from Catnic Components anchor the wall plate to the wall, and another form of strapping is available which holds the joist to the wall. These are shown in the accompanying illustration (*see* Fig 16). Fixing joists to the wall plates is done by cross-nailing – driving two 100mm (4in) wire nails through the joist from either side, and into the plate (*see* Fig 17). On a steep slope, a joint needs to be cut in the joist known as a 'birdsmouth'. This is described in respect of rafters and pitched roofs in Chapter 4.

Completing structural support and fascias

It is important that joists are trimmed to an *exact* length in order to provide an aligned structure for receiving the fascia boards. The way to do this is to mark cutting points with a string line as shown in the accompanying photo sequence (Photos 32a, 32b and 32c). Having marked where the line crosses the timber, a vertical is dropped with a spirit level to give an accurate sawing line.

Eaves and verge detailing can be constructed in several ways. The amount of overhang is a matter of preference, and in order to make a provision for ventilation apertures, the practice formerly adopted of fixing fascias direct to the brickwork should be avoided. Cantilever rafters should be added,

Photo 32a Joists supporting a deck for a flat roof, or rafters forming a pitched roof are trimmed to length in the same way. These photographs show the marking-out procedure. The projection required for the soffit and fascia boarding is measured out from the face of the building

Photo 32b A string line is stretched tautly and pencil markings added so that the cutting point on each projecting timber is marked in exact alignment

Photo 32c Using a spirit level, a perpendicular is dropped from each point where the string intersects with the rafter or joist. A vertical cutting line is then drawn in to help with accurate sawing

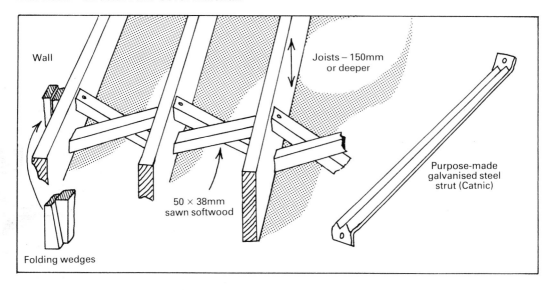

Fig 18 To prevent the deflection of deep joists, herringbone strutting and folding wedges must be added. 'Catnic' steel struts are an alternative to timber herringbones, and these are available in a variety of sizes

Fig 19 Decking in either exterior grade plywood or high density particle board is fixed to the roof structure with lost head round wire nails. If the framework includes outer rafters, these improve its structural integrity, and offer a sound base for the fascia and soffit boards which are added next

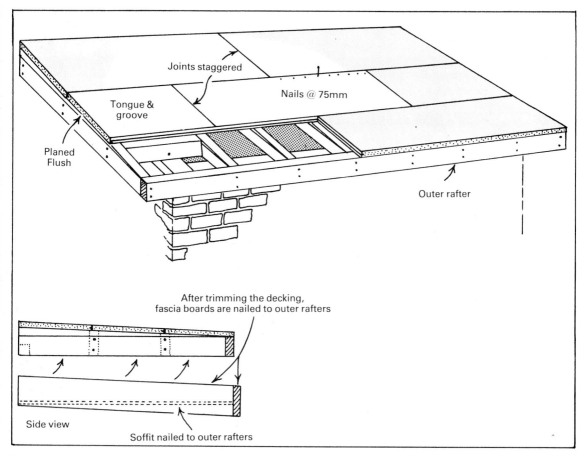

marked with the string line technique and trimmed to length. To produce an overhanging deck, these are merely cantilevered over their support wall (*see* Fig 14, page 51).

Where deep joists of 150mm (6in) or more are used, bracing is needed to give lateral support and to prevent twisting. Traditional herring-bone strutting from 50 × 50mm or 50 × 38mm (2 × 2in or 2 × 1½in) sawn timber achieves this objective while preserving a ventilation airway. Cutting each strut at an angle may take the beginner a little time to execute accurately, and an alternative is to use purpose-made galvanised steel struts (*see* Fig 18). Catnic Components make struts to suit joists of four heights from 150mm (6in) to 225mm (9in), and at the three spacings already described on page 46.

When joists have been cut to length, fascia boarding can be nailed directly to the exposed ends. A fascia board is typically formed from 25mm (1in) planed all round (*PAR*) softwood, which, with the loss of material in machining, bears a finished thickness of around 22mm (⅞in). This deflects easily, and a sturdier arrangement is to bridge the ends of joists or extension timbers with outer rafters of a dimension similar to the joists themselves as shown in Fig 19. This additional outer rafter provides better support for the fascia, responds less easily to deflection, and contributes to a much sturdier deck structure.

Fascia board should be fixed with round-headed wire nails which have been galvanised. The heads should be driven below the surface with a nail punch, and the recess filled with an exterior-grade wood filler. Take care not to bruise the surface of the board with the head of the hammer; invariably these dents are unpleasantly conspicuous – particularly when the fascia has been painted.

Decking

At one time, close-boarded or tongue and grooved boards were used for roof decking, but nowadays these are regarded as unsatisfactory. In modern practice, water-resistant plywood is recommended, and this is often designated by timber suppliers as 'WBP exterior grade plywood'. Clearly this is excel-

lent for its task, but many builders prefer to use chipboard which is cheaper, easy to work and reasonably stable (ie it is not prone to shrink or expand in changing weather). This deck material is not approved in Codes of Practice, and if it becomes damp it may start to sag. Nevertheless it is often used for porches and garages, and if you decide to use chipboard it is most important to specify a high-density or 'roof deck-grade chipboard'. Moreover, this should be purchased with tongue and groove (T & G) edges so that joins made across the joists require no support noggings underneath. This is readily available and 2440 × 600mm (8ft × 2ft nominal) sheets are stock items at most timber merchants. It is unfortunate that high-density particle board soon blunts saws, but it is easy to use. Joins in adjacent runs of boarding should be staggered, and you shouldn't skimp on the nailing. Lost head round wire nails are ideal, and fixings should be made at intervals no greater than 75mm (3in) (*see* Fig 19). A wooden spacer piece cut to 75mm (3in) is a useful help-mate. Additional strength is provided if woodworking adhesive is squeezed into the T & G joint prior to forming the interlock; I always follow this practice – although few professionals bother. It is essential to keep the material absolutely dry in readiness for its built-up felt covering and you must have a tarpaulin ready to cover up the work areas. If there is any likelihood of sudden showers, you might prefer to purchase pre-felted chipboard which gives an additional measure of protection.

Without question, chipboard is second-best to WBP plywood. However, with current concerns about energy conservation and growing interest in 'warm-roof' construction, it is significant that composite boarding is now available in which an insulant foam is bonded to exterior-grade plywood. This offers an alternative approach to decking.

An example is Coolag 'Purldek' in which polyurethane foam is bonded to 8mm (⁵⁄₁₆in) exterior-grade ply. The material also features an aluminium foil to complete the three-component sandwich. This means that a single product provides a foil vapour barrier, an insulant and a plywood sheet decking. For the vapour barrier to be effective it must be continuous – so joints in the board have to be sealed with a mastic during construction. Inevitably, composite decking materials are more expensive, but their special appeal is the fact that they combine three functions in one, and installation time is thus reduced.

Equally innovatory is Spandoboard which is an insulant made with a tapering profile to install on an existing deck. It is made with a polyurethane or polyisocyanurate insulating foam, sandwiched between 12.5mm (½in) bitumen-impregnated fibre boards. A vapour barrier membrane must be added during installation, and the boards are specially made to produce falls of 1:60 or 1:80. An ordinary handsaw is all that is required to cut the boards to size.

Of special interest to the amateur builder is

the fact that new decking materials like Purldek and Spandoboard are easy to use. Products, initially introduced for large roofing contracts and industrial installations, are now being used more and more in domestic building. On existing structures with 'cold roofs', the need to add ventilation must undoubtedly be recognised. Meantime, the arrival of integrated insulation deck materials has generated considerable enthusiasm for 'warm-roof' designs.

Drips and upstands

Before a built-up felt covering is finally added, preparatory work is needed at the eaves, verges and upstands. Rainwater should discharge from the lowest point, known as the eaves, and it is important that the roof boundary extends beyond the fascia board so that any flow is shed beyond timbers and brickwork. The structure should also align with a run of eaves guttering so that the

Photo 33 Multi-function roof decking, which comprises a high efficiency insulant and vapour barrier sandwiched between plywood, is ideal for the DIY builder. This construction system overcomes the condensation problem in flat roofs

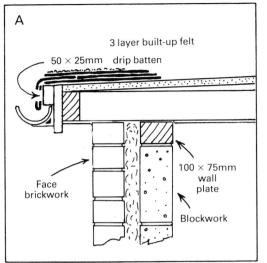

A

3 layer built-up felt

50 × 25mm drip batten

100 × 75mm wall plate

Face brickwork

Blockwork

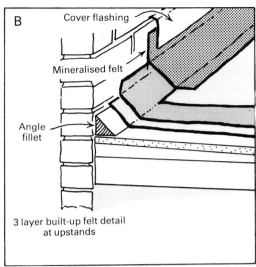

B

Cover flashing

Mineralised felt

Angle fillet

3 layer built-up felt detail at upstands

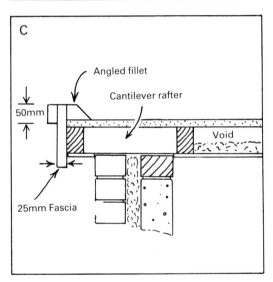

C

Angled fillet

Cantilever rafter

50mm

Void

25mm Fascia

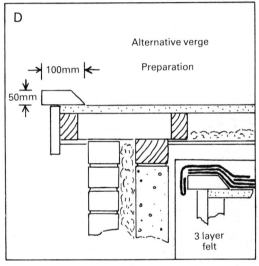

D

Alternative verge Preparation

100mm

50mm

3 layer felt

Fig 20 Prior to felting a flat roof, the decking must be prepared at verges, eaves, and abutments with drip battens and fillets

A A 50 × 25mm drip batten is needed to support the drip skirt at the eaves, ie the low point of the roof
B At upstands, a timber angle fillet must be nailed into place. The subsequent build-up of felt, and the use of a cover flashing is shown in the diagram
C At the verges, ie the sides of the roof, a popular strategy is to extend the fascia board above the decking, against which is mounted a drip batten and an angled length of fillet timber
D As an alternative to C, verge preparation can be carried out in a single operation using 100 × 50mm timber. This must be prepared with a circular saw set in the tilt mode to produce a bevel edge

rainwater is collected efficiently. As shown in Figs 20a-d, this extension is achieved by nailing a 50 × 25mm (2 × 1in) batten to the fascia board. When a welted apron of mineralised felt is subsequently added, it will be fixed to this batten with galvanised clout nails, coated with bitumen and folded back on itself. By extending the felt strip *beyond* the lower edge of the drip batten, the apron will discharge water directly into the guttering, and drips will be released rather than drawn towards the fascia by capillary action.

At the sides of the roof, referred to as verges, the finish can be achieved in several ways. A typical detailing is shown in Fig 20c, in which an angled fillet has been nailed to the decking. This detailing is known as a 'water check', and helps to contain water within the roof area rather than spilling from the sides. A 50 × 25mm (2 × 1in) drip batten is again needed, and a strip of welted felt, as described for the eaves, is added later.

On a lean-to flat roof, the upstand at the abutment wall needs some preparatory work to reduce the sharp angle. This is done by adding an angle fillet along the point of intersection between the deck and the vertical abutment (*see* Fig 20b). You will also need to prepare the wall abutment to accept a vertical upstand of at least 150mm (6in). With standard-size bricks, this represents two courses of brickwork, above which the horizontal mortar joint should be deepened with a plugging chisel. This gives a recess into which the felt can be tucked and anchored. To improve the weatherproofing, a cover apron of lead or adhesive foil flashing should also be added, a procedure discussed in Chapter 7.

Alternative verge details

It is also possible to finish side details using purpose-made edgings. These are available in reinforced plastics or non-ferrous metals like aluminium, and take the place of a welted felt drip (*see* Fig 21). Sections are

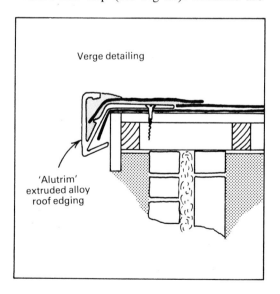

Verge detailing

'Alutrim' extruded alloy roof edging

Fig 21 As an alternative to timber fillets, flat roofs can be edged with trims moulded in plastic or aluminium. This example, the 'Type F', is designed specifically for felt roofs – one of nine in the 'Alutrim' range of extruded alloy edgings

fixed to the decking with woodscrews. Products such as 'Alutrim' are made in a variety of profiles to give an efficient water check at the verges as well as an attractive finish. Fixing procedures and installation diagrams are provided by the manufacturers.

Built-up felt coverings

The installation of overlapping layers of roofing felt is a popular way to create a continuous cover on a flat roof. The layers need to be fused together when installed, and the traditional way to do this is to use hot bitumen. This is prepared in a bitumen boiler, taken to the roof in a bucket, and poured accurately so that the felt can be unrolled directly over the hot coating. The procedure is descriptively called the 'pour and roll' technique, and Photo 29, (page 44) shows it clearly. A more recent development uses special pre-coated felts, which are mounted on to a roller dispenser and heated with a propane-gas heat torch. As the flame from the torch is played on the felt, the bonding agent is softened and activated while the material is rolled into position (*see* Photo 30, page 44). This system obviates the need for a bitumen boiler, although it calls for good workmanship to produce sound adhesion – especially around the edges and at overlap points.

The pour and roll technique is a skilful operation, and boiling bitumen is potentially very dangerous. It is taboo in the trade even to ask a contractor if he has ever been burnt; to talk about a slip on a ladder when carrying a bucket of bitumen is never discussed. If hot bitumen is poured on to a damp surface, the reaction is not dissimilar from splashing water into a hot chip pan. The spitting can easily cause facial injury or burns on the hands and arms. It is fully recommended to get this work done professionally. Notwithstanding this advice, it is interesting to see the vigorous promotion of torch-on techniques as an alternative to pour and roll roofing. Moreover, the manufacturers of Permanite 'Britorch' regard this method of application as an operation which is well within the capacity of competent amateurs. A gas bottle and roofing torch will need to be hired, and it would be wise to start on a small

area such as a tool shed before tackling a large roof. But if this is acknowledged, the task is worth contemplating.

Even if you decide to appoint a roofing contractor, it is useful to know something about the operations which he should carry out. There are a number of companies which manufacture felt materials, but in general you will find that they are remarkably good at providing information on their products, and in explaining the way in which they should be installed. For instance, the free booklet *Built-up Roofing for the Builder* by Ruberoid has been well known for many years, and a revised edition was updated in 1985. Reference has already been made in the previous chapter to Permanite's *Fixing Bitumen Roofings,* which describes cold-bond installations. Anderson Roofing has similarly launched packs directed at DIY roofers, and excellent back-up literature like *Built-up Roofing in the UK.* Accordingly, this section is kept brief, and if you require more detailed information, you will find manufacturers extremely helpful.

Photo 34 The second of three layers of standard fibre reinforced felt on a dormer window, being carefully bonded and taken over the top of the ridge board

Materials

It must be stressed that 'roofing felt' is a title for many different materials. In particular, you must not confuse the felts used for built-up felt roofs with 'sarking felt', which is laid below slates or tiles. Many different types are described in a British Standards document entitled *Specification for Roof Felt* (BS 747 : 1977). Unfortunately, the newer 'high-performance' materials which have been developed in the last few years are not listed, but many have been tested and approved by the British Board of Agrément. Meantime, a revision currently being made to BS 747 is expected to include Hyparoof High Performance membrane: it may also be presumed that other products will also be added in due course.

When constructing a covering, the installation begins with one or two 'base layers'. In a three-layer system, BS 747 recognises three grades of base layer felts which are differentiated by their reinforcing fabric:

Class 1 – fibre reinforcing which is very flexible and low in cost.

Class 2 – asbestos reinforcing which is noteworthy in respect of fire resistance, and less likely to rot.

Class 3 – fibreglass reinforcing material which is used for high-quality work and is completely rot-proof.

Roof coverings which comply with BS747 are relatively cheap, but lack the tensile strength and life expectancy of coverings formed with 'high-performance' materials. In the trade, the 'high-performance' designation usually refers to a double-layer system using materials which feature a polyester base, for example Permanite 'Hyparoof'. There is good sense in specifying a high-performance material since its glass-fibre base is both inert and rot-proof. When you compare literature from manufacturers, you will also come across 'polymeric systems' in which the felt has inherent strength without a reinforcing material. This is designed for high-quality applications which will be installed by a professional roofer.

On domestic flat roofs, the cheaper materials complying with BS 747 have been the most popular choice, but this is likely to change when the benefits of high-performance products are recognised. In choosing a material there are many factors to consider, such as roof shape, whether the structure is a 'hot' or a 'cold' roof, the material used for the decking, the available budget and the life expectancy required. Different felts matched with different bonding agents contribute to a multitude of alternatives, and one of the best guides for selecting products appears in *Built-up Roofing in the U.K.* available from Anderson Roofing.

After the base layers (or layer), the top layer is referred to as the 'cap' or 'capping sheet'. To comply with the Building Regulations, some form of solar protection is also required. Special felts manufactured for this purpose are sometimes used, and these may incorporate a reflective top layer such as aluminium foil. Another strategy is to coat a conventional mineral-surfaced felt with a solar-reflecting paint, such as 'Solabar' (Anderson Roofing). However, a common practice is to place a final cover layer of 12.5mm (½in) maximum size limestone or granite chippings. These contribute to fire resistance as well as giving solar reflectivity. Less endearing is the tendency for the chippings to dislodge and fall into eaves gutters, a problem which can be avoided if a mineral-surfaced felt is used. The highest fire-rating standards are achieved by many mineralised cap sheets.

Installation

Procedures for installation vary according to deck material and the choice of felts. On a timber deck, the first layer is usually nailed in place with clout nails; this permits a degree of movement in the event of differential expansion between the deck and cover materials. Further layers, however, must be bonded with bitumen. Installation always commences at the bottom of the slope and works upwards, and each layer will be placed so that joins never coincide. To obtain maximum weatherproofing, the best approach is to place roll lengths *parallel* to the fall line. At the sides, adjacent pieces of felt should overlap at least 50mm (2in) – referred to as 'side lap'. At the ends of rolls, a minimum 75mm (3in) overlap is necessary – referred to as either 'head lap' or 'end lap'. However, on roofs with a slope of less than 10 deg, head lap should be increased to 150mm (6in). Where high-performance felt is used, the respective minima are 75mm (3in) side lap and 150mm (6in) end lap.

Around the perimeters of the roof a mineralised felt is used to form the welted drips and upstands. This is installed in strips, and the technique for producing wraparounds at corners and joins is illustrated in manufacturers' guides. Strips are nailed to the drip battens and, by folding the material back on itself in a welt, the heads of fixings are hidden.

Roof performance

Whereas the unit coverings of pitched roofs, such as concrete tiles, are guaranteed (Redland tiles are guaranteed for a hundred years), no such assurances are given with built-up felt roof materials. In some respects this is understandable; long-term performance is dependent on many features which fall beyond the control of the cover manufacturer. The stability and integrity of the substructure is one important contributor to longevity; the workmanship of the felter is another. Whereas a high-performance roof covering might give twenty-five years of maintenance-free service, a rag-fibre-based covering may need replacement after a decade. However, there is an element of caution inherent in statements about longevity. High-performance membranes have now been available for fifteen years, and on the evidence of performance to date manufacturers are confidently expecting a life expectancy of at least forty years. To help meet this objective, regular inspection is strongly advocated, but many home owners overlook this chore – until rainwater starts to leak through the ceilings.

In consideration of problems, cracks in the covering are often the result of deflection in the substructure. Unavoidable settlement or thermal movements place stress on the felt layers and this can lead to fractures. Extremes of surface temperatures caused by the sun constitute a similar challenge to the covering. Whereas slates or tiles accommodate movements through their independence, a continuous cover material of roofing felt, which may occupy a large area without a break, is subject to particular stress. In this connection, it is important that the means of solar protection is maintained in good condition. This is also a reason for recommending the polyester-based materials with polymer coatings as these are designed to accommodate movement without sustaining damage.

Fig 22 A terrace or walkway can be constructed on a built-up roof of sufficient strength using slabs and Euroroof's 'Bishore' paving supports. The supports distribute loads efficiently, provide suitable clearance for surface drainage beneath the slabs, and are easy to lay

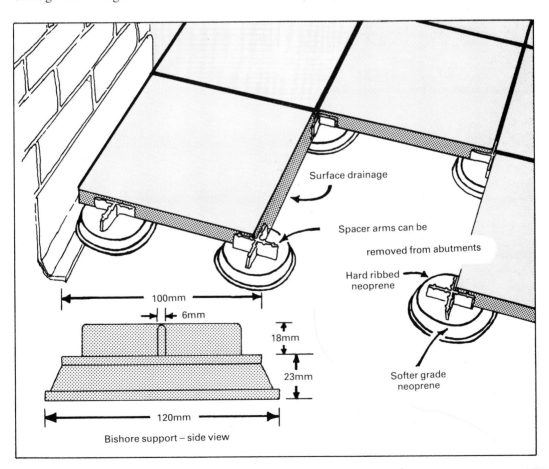

Surface drainage

Spacer arms can be

removed from abutments

Hard ribbed neoprene

Softer grade neoprene

100mm

6mm

18mm

23mm

120mm

Bishore support – side view

The slow rate of rainwater flow on a flat roof is another contributor to failure, and weaknesses are highlighted at an early stage. It is all-too-easy for moisture to creep under deteriorating joins by capillary action. Notwithstanding these problems, temporary repairs can often prolong life considerably, and a number of chemical coatings are available. It is also possible to add a new cover *on top* of the existing one, and refurbishment advice is included in *A Guide to Flat Roof Maintenance and Refurbishment* published in 1985 by Euroroof Ltd.

Balcony tiles

Before leaving flat roofs, it should be mentioned that tile and paving arrangements can sometimes be used on deckings for sun roofs and high-level patios. Balconies are undoubtedly popular, but it must be repeated that flat roofs cannot be used as sun decks if their structure was designed for occasional-repair accesss only. However, if there is no doubt about the integrity of the support structure, walkways can be constructed. Euroroof are specialists in various forms of roof paving, and their 'Roofpave' non-slip glass-reinforced tiles are specially made to bed into hot bitumen. In addition to providing a safe surface, they also give solar protection to the cover material. Also of interest to the amateur builder are Bishore paving supports from Euroroof. As shown in Fig 22 these allow you to install a covering without special tools, and as long as the substructure is sound they will even support paving slabs. However, in most domestic situations, verandas and balconies are constructed with smaller and lighter tiles. A special feature of the Bishore system is the fact that drainage occurs *below* the surface of the units, and since the principle of installation involves no bonding, it is easy to lift tiles or slabs to inspect the structure below.

4 Construction of Pitched-roof Structures

Self-build enthusiasts with competence in carpentry often tackle the construction of simple pitched roofs without too much difficulty. To do this successfully, it is presumed that roof design and specification have been undertaken by an architect or building surveyor. Equally, it is important that detailed drawings have been prepared which provide the information relating to construction. For whereas a competent amateur might successfully build a pitched roof, he should not be misled into thinking that he could design one. It has already been pointed out that to calculate the strength of a roof structure, a designer must take into account many factors including wind effects, snow loadings and so on. It is also important for the beginner

builder to recognise the wisdom of tackling a pitched roof of *simple* design. Unusual examples like roofs featuring sprocketed eaves, or roofs with complicated features such as eyebrow dormer windows are best left to experts. By the same token, a roof featuring hips and valleys will involve a higher level of joinery skills than a straightforward duo-pitch structure.

Another matter of relevance concerns the form of construction. In domestic building, there has been a marked swing away from the traditional approach where purlins and raf-

Photo 35 To construct the main roof with pre-formed trusses took a day to complete; garage and porch built traditionally took a fortnight. Such is the case favouring modern methods

ters are cut and fitted on site. The use of pre-formed trusses has simplified construction and speeds up the building process. In my most recent self-build project, this time saving was underlined most clearly. As shown in Photo 35, the porch and double garage were built traditionally, whereas the main roof was constructed with roof trusses. Working single-handed, and with the appropriate cautionary approach of the DIY builder, the installation of purlins and the jointing and fixing of rafters took two weeks to complete. In complete contrast, the main roof built with trussed rafters took no more than a day, although a helper was needed to lift and manoeuvre the units into position. This demonstrated in emphatic style why trussed roofs are a familiar sight on almost all volume-construction housing estates. Indeed it has been claimed that over ninety per cent of the roofs on all new domestic premises in this country are constructed with prefabricated trusses.

Of course there are a number of other reasons why trussed rafters have achieved this position of prominence in modern building. The fact that each unit is fabricated identically contributes to a symmetrical structure, and the fact that its design is established by computerised calculation assures sound integrity using the minimum of timber. It is worth adding, too, that scarcely any reliance is placed on a carpenter's ability to form sound joints. If you use manufactured units, building a roof can be seen more as an assembly exercise. It goes without saying that certain procedures must be strictly observed if the system is to succeed, but for the amateur builder trussed systems are ideal. But this chapter is not concerned solely with this form of construction for two reasons.

Firstly, if a roof space is needed for a loft conversion project later on, most pre-formed roof trusses preclude this development. It is true that special attic trusses can be designed and manufactured to order, but most conversions proceed in roofs with traditional purlin and rafter construction. Indeed it was for this reason that my self-build house, shown alongside (*see* Photo 35), featured a traditional roof over the garage when trussed rafters were used for the main roof. A cry for more space was easily satisfied, and building

an extra room over the garage was straightforward. But another reason to look at traditional techniques concerns refurbishment work. The nature of many older properties usually commends 'on-site' construction rather than prefabrication. For example, the structure of an 'elderly' country cottage might not be satisfactorily met by manufactured trusses, nor indeed would the building's individual charm be enhanced by their geometrical exactitude. The contrasting features of new and old buildings call for different treatments.

Preparing wall plates

Irrespective of constructional method, roof timbers or trusses need to be fixed to a timber wall plate. The function and fixing of a 100 × 75mm (4 × 3in) plate have already been described in Chapter 3, and you should check these points. In modern building practice, it is usually necessary for the plate to be bedded in mortar, *and* mechanically fixed to the wall with purpose-made metal straps. This provision may be unnecessary if a heavy tile is specified, but the decision is best left to the truss designer who will make calculations on the basis of the tiles chosen, wind speeds, exposure and building height.

Prefabricated trusses

In traditional forms of roof construction, support was often available from load-bearing internal walls. Today's modern houses are built differently, and quite often the only means of support are the exterior walls; at levels above the ground floor, internal walls are generally constructed merely to divide living spaces, and cannot be used to give structural support to the roof. Fortunately, however, a prefabricated roof truss is able to carry loads over large spans. Trusses can thus be positioned on wall plates fixed to the inner skin of a cavity wall – it is the inside 'leaf' of a cavity wall which usually carries the roof load rather than the outer skin.

Truss technology has made notable advances during the last two decades, although it is interesting to observe that traditional methods still prevail in many countries and regions abroad. For example, roof structures in Normandy and Brittany are still principally constructed with purlins, rafters and close boarding. However, one should not underestimate the strength of prefabricated trusses and the impressive integrity provided by their galvanised punched metal-plate fasteners. These join abutting timbers, and dispense with the need to form interlocking joints – cutting a joint inevitably introduces a weakness at points where the cross-section of timbers is reduced.

Notwithstanding any strength/weight benefits, the design, construction and assembly of trusses must *not* be attempted by the amateur builder. Rigorous requirements are laid down in British Standards (BS 5268, Part 3), and manufacturing techniques are sophis-

ticated. Hydraulic presses are used in the fabrication process, and special jigs ensure identical repeats. But, whereas the finished product has remarkable strength as a roof support, careful site handling and on-site storage is essential. Moreover, the timbers should not be drilled or cut, and joining plates must not be damaged.

A number of truss designs are in common use, and these can be configured to suit mono pitch, duo pitch and asymmetric roofs (*see*

Fig 23 Examples of trussed rafter configurations to suit different types of roof design

Fig 24 Fink trussed rafter showing proportions, components, and the pitch/span details which are required by a designer/fabricator when an order is placed

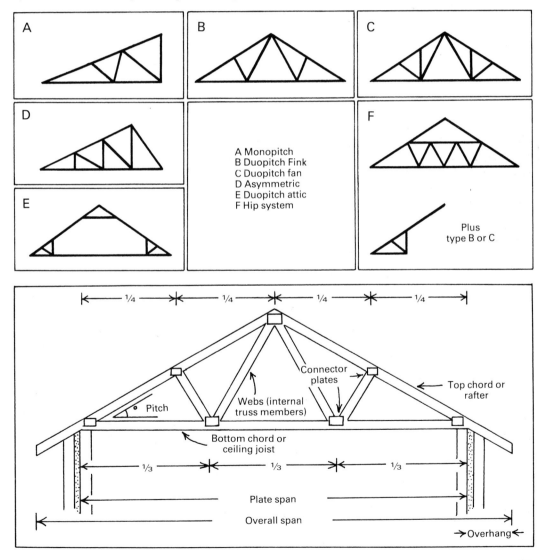

A Monopitch
B Duopitch Fink
C Duopitch fan
D Asymmetric
E Duopitch attic
F Hip system

Plus type B or C

Fig 23). You are most likely to come across the 'fink' truss in domestic buildings, although 'fan' trusses are sometimes specified. As shown in Fig 24, the fink truss with its familiar W pattern of webs is designed to divide the span between wall plates into three equidistant measures on the bottom chord.

Ordering

An architect's drawing will indicate the trusses required, and these can be ordered from a specialist timber merchant. A number of details will be needed, and order forms usually ask the following questions:

1. Number of trusses required.
2. Truss centres – ie spacing from the centre of one truss to the centre of the next.
3. The plate span – taking the measurement from the *outside* edges of the wall plates (*see* Fig 24).
4. Overhang at the eaves, measured from the outside of the wall plate (*see* Fig 24).
5. Overall span from the extremities of the truss (*see* Fig 24).

Fig 25 Ideally trusses should be stored in the upright position and stacking should be carried out against a firm and safe support. The trusses should be supported at the design bearing points and at such a height as to ensure that any rafter overhangs clear the ground

6. Pitch (or pitches) of the roof.
7. Dead loads to be supported by the rafter as stated in the specification data of the architect's drawings. (Calculated in accordance with the covering units, and using information from the tile manufacturer.)
8. Preservation treatment needed (if any).
9. Position and capacity of water tank(s).
10. Any special loadings extra to those quoted.

Storage

Wherever possible, trusses should be stored upright on a pair of timber bearers as shown in Fig 25. Some builders choose to store them on their sides, but this requires many more bearers and can lead to distortion if support is poor (*see* Fig 26). In either case, the storage area should be free from vegetation, and trusses must be covered to give complete weather protection.

Handling

Wherever possible, trusses should be carried upright – in the plane which they will assume when placed on the wall plates – and it is best if they can be held at the eaves joints. They

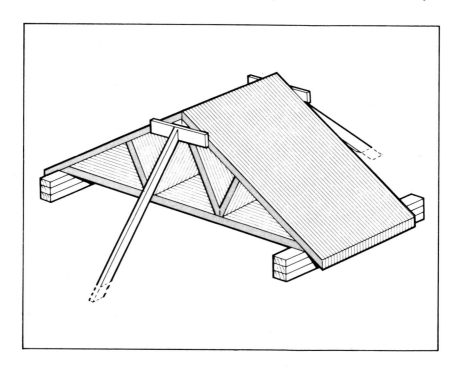

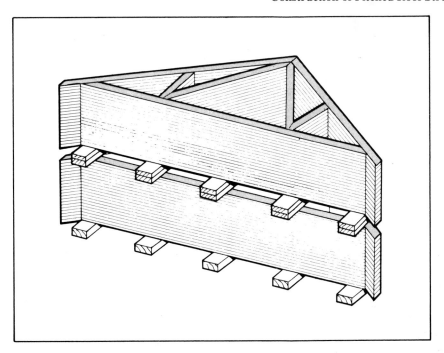

Fig 26 Where trusses are to be stored horizontally, bearers should be placed to give level support at close centres, sufficient to prevent long term deformation of all truss members. If subsequent bearers are placed on top of other trusses, they should be vertically in line with those underneath

Fig 27 Where single trusses are being moved they should be carried with the ridge down unless sufficient labour is available to provide full support

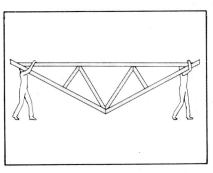

can be carried in an inverted position (*see* Fig 27), but you must exercise great caution when turning a truss over. Lifting trusses into position will undoubtedly need helpers (*see* Fig 28) and to move them carelessly may lead to irreparable damage. For instance, if man-handled sideways the truss is placed under severe stress; if see-sawed over a wall or scaffold walkway, the joints may fracture (*see* Fig 29). Don't risk costly damage; draw together a full complement of assistants. With help, a complete set of trusses can be lifted up a scaffolding to roof level in minutes rather than hours.

Supplementary bracing

When a roof is constructed with trussed rafters, bracing timbers have to be added both lengthways and diagonally. This provision was the subject of amendment in the revision of BS 5268 : Part 3 : 1985; if you order trusses, you must consult the manufacturer for up-to-date guidance.

Longitudinal bracing is carried out with timbers measuring 100 × 25mm (4 × 1in), and pairs of galvanised wire nails 75mm (3in) long, and 3.35mm (⅛in) diameter (10 gauge). If shorter lengths of bracing timber are used, these must be lapped over a minimum of two trussed rafters and nailed. Given spans no greater than 8m (26ft 3in), five longitudinal binders are needed for a roof structure made with fink trusses (*see* Fig 30). On a duo-pitch roof, these should extend beyond the last truss to butt tightly against gable walls. Diagonal rafter bracing is also required using 100 × 25mm (4 × 1in) timber

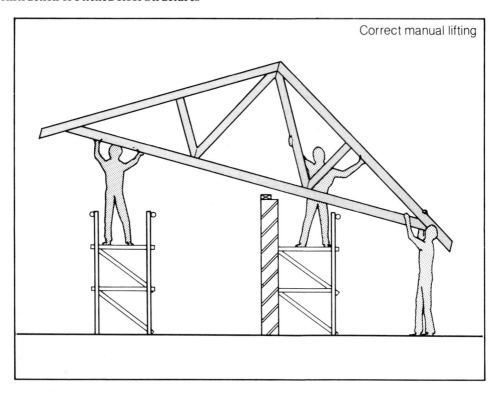

Correct manual lifting

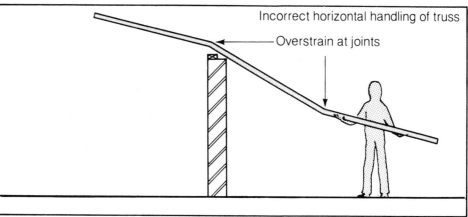

Incorrect horizontal handling of truss

Overstrain at joints

Fig 28 On long span trusses it may be necessary to employ additional labour at intermediate positions. If required the truss may be inverted so that the apex hangs down. See-sawing the truss across walls and scaffolding must be avoided. Individual house designs and site conditions may dictate different requirements in order to install trusses in their final position. Consideration of the method to be used should be given before lifting begins

Fig 29 Incorrect method of lifting trusses into position. Normally the greatest stresses which truss joints will undergo are those caused by site handling and lifting. It is important that wherever possible the trusses should be lifted in the vertical plane, apex uppermost. Handling should always be carried out with the utmost care to avoid possible damage to both timber and connector plates

fixed with pairs of nails 63mm (2½in) and 3.35mm (⅛in) diameter, 10 gauge, as shown in Fig 30. At the peak of these diagonals, a 45 degree angle should appear between the brace and the line of the ridge when viewed from above or below. The truss where this brace terminates represents the first unit to be erected on a roof. (Truss 'A' in Fig 31).

Over an unusually large span, where the supporting walls are spaced more than 8m (26ft 3in), you will need special additional bracings which the truss manufacturer will describe on request.

Guidance for erecting trussed rafters

Several manufacturers produce basically similar types of punched metal-plate fasteners on which the prefabrication of roof trusses is dependent. These companies are eager to promote good building practice so that the performance of their product is fully recognised. To implement this objective, detailed literature is produced by many of the system manufacturers to describe constructional procedures. For example, the booklet produced by Gang-Nail Ltd, entitled *Trussed Rafter Construction and Specification Guide* clearly illustrates and explains subjects like negotiating awkward chimney openings, or forming access hatch openings which cannot be situated between truss positions. Hydro Air International (UK) Ltd provides a similar service with data sheets covering topics such as the construction of supports for water tanks. The information provided here and reproduced with the permission of Gang-Nail Ltd, provides general guidance for the self-builder, but for more detailed information manufacturers' literature should be consulted. Alternatively, you will find useful advice contained in the *Technical Handbook* from the International Truss Plate Association. Addresses are given in the Appendix.

Erection procedure on a duo-pitch roof

Step-by-step procedures on a roof whose span does not exceed 8m (26ft 3in) should be read in conjunction with Fig 30. As mentioned earlier, work commences by installing the truss whose situation falls at the head of a corner diagonal brace.

Fig 30 Key to the step-by-step erection procedure described in the text. Truss (A) at the head of a diagonal brace is the first to be fixed, and held plumb with a temporary strut (B). Note the use of permanent longitudinal bracing timbers (G) and diagonal bracing (F). Temporary bracing (D) gives additional integrity to the structure as an interim measure before the later installation of tile battens

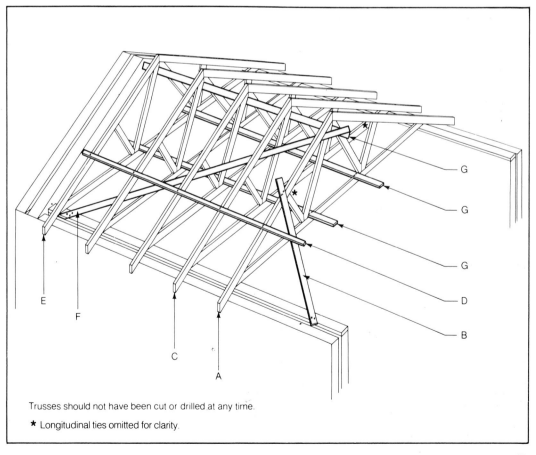

Trusses should not have been cut or drilled at any time.

★ Longitudinal ties omitted for clarity.

Construction of Pitched Roof Structures

1. The centres for each truss should be measured and marked on the wall plate in accordance with the architect's drawings. Centres are usually spaced at 600mm (24in) intervals. (Ceiling boards are manufactured in sizes which assume this spacing.)

2. The first truss (A in Fig 30) is offered up to position and held with several temporary braces (B) nailed to the top of rafter sections and the wall plate. It must be checked with a spirit level or plumb line to ensure that it is absolutely vertical. Measurements must also be carefully taken to confirm that the truss is parallel with the gable walls.

3. Trusses can either be fixed to the wall plate using purpose-made truss clips (*see* Fig 31), or by skew nailing. In a region noted for high winds or on an exposed site, truss clips should be used as a matter of course. Do not fix the clips fully until all trusses are in position and checked for accuracy. In sheltered situations skew nailing is acceptable, which involves driving a pair of nails angled from opposite sides through the truss and into the wall plate. If necessary, nails can be driven through slots in the heel connector plate, but make sure that the nails are not driven fully

Fig 31 Purpose-made truss clips give accurate and effective anchorage. However, a competent carpenter may prefer skew nailing as an alternative method for fixing trusses to the wall plates

home – the head should remain clear of the connector plate to avoid distortion. In addition, care must be taken to ensure that the angle is sufficiently acute to avoid any risk of dislodging the connector plate on the opposite face of the truss. Galvanised round wire nails – 100mm (4in) long and 7 gauge – are recommended. Good skew fixings require a certain amount of skill, and the beginner is advised to use truss clips instead.

4. Trusses are subsequently added towards the gable-end wall (starting with C and ending with E in Fig 30), and linked to the first with temporary longitudinal bracing timbers (D in Fig 30). A useful tip is to cut your own spacer gauge from an offcut of timber to cut down on the use of the measuring tape. The length of the gauge will equal the centre spacing of the trusses *minus* one truss thickness. Offering it up to adjacent trusses will quickly confirm correct spacing. As the trusses are erected you are advised to nail additional temporary members on top of each rafter. These ensure that the spacing at the wall plate is preserved at the peak of the trusses. When tiling battens are added later, these make an important contribution to structure, and must be of sufficient strength to achieve this objective (*see* pages 112-13).

5. The stability of the structure is achieved as soon as the trusses are tied in with each other.

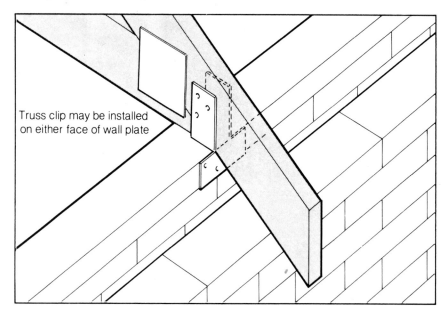

Truss clip may be installed on either face of wall plate

70

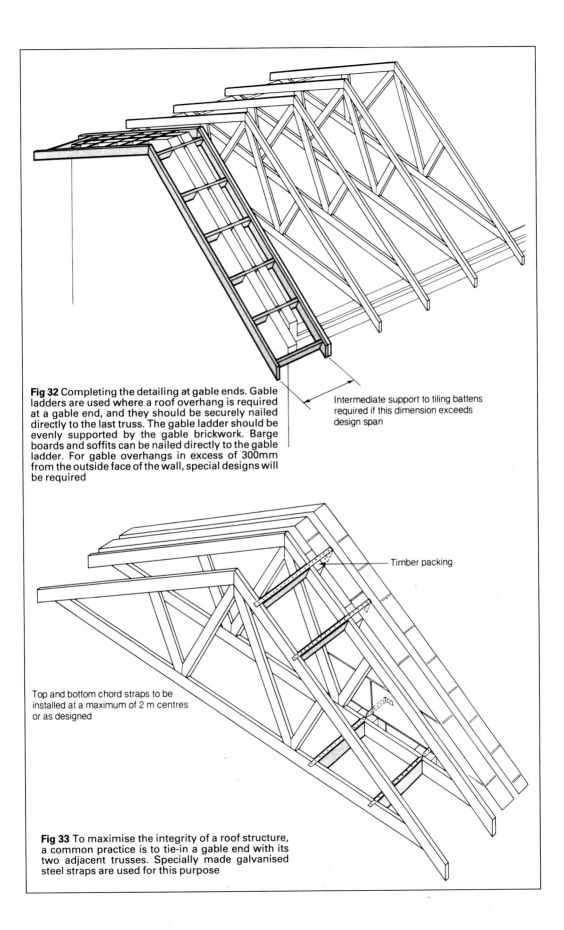

Fig 32 Completing the detailing at gable ends. Gable ladders are used where a roof overhang is required at a gable end, and they should be securely nailed directly to the last truss. The gable ladder should be evenly supported by the gable brickwork. Barge boards and soffits can be nailed directly to the gable ladder. For gable overhangs in excess of 300mm from the outside face of the wall, special designs will be required

Intermediate support to tiling battens required if this dimension exceeds design span

Timber packing

Top and bottom chord straps to be installed at a maximum of 2 m centres or as designed

Fig 33 To maximise the integrity of a roof structure, a common practice is to tie-in a gable end with its two adjacent trusses. Specially made galvanised steel straps are used for this purpose

Photo 36 Constructing a roof in trussed rafters in this detached double garage represents an ideal first-time project for the amateur builder

At the earliest opportunity, permanent diagonal bracing (F in Fig 30) should be nailed to the underside of all the rafters. This is supplemented by the binding effect of the longitudinal timbers (G in Fig 30). Round wire nails are recommended for all fixings.

Where the run is longer than the timbers available for bracing, the joining point should have a lap (ie doubling up) which extends over at least two trusses, with double nailing at each brace/truss intersection on both pieces. This *is* important even though you may notice that many professional builders seem content not to heed the instructions of the truss manufacturers.

Completing the detailing at gable ends is shown in Fig 32 and 33 and is similar to the procedures involved with traditional roofs

(*see* pages 82-3). As far as the amateur self-builder is concerned, the advent of trussed systems enables him to tackle a roof with confidence. If it is possible to commence with a small project such as the detached garage shown in Photo 36, so much the better.

Water tanks

If the size and position of an attic tank has been given to the truss manufacturer, units will have been manufactured specially to bear this very considerable loading. Tanks up to 200L (44gal) have to be supported on a base which bears on three trusses. Larger tanks up to 300L (66gal) require a weight distribution over four trusses. Designs for the platform based on the recommendations for Gang-Nail products are given in Figs 34 and 35. Where headroom is a problem (in the case of low-pitched roofs), the height of the platform can be reduced by using joist hangers to support cross-timbers.

72

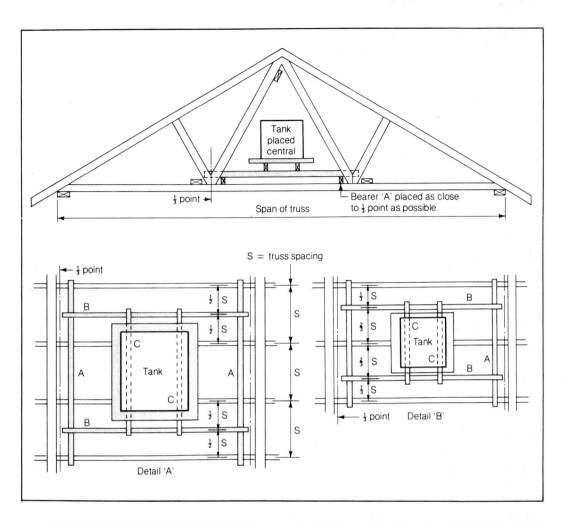

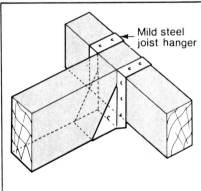

Alternative support between members

Where space is limited this detail may be used between members A & B and B & C in order to gain head room. However a minimum clearance of 25 mm above the ceiling lining should be allowed, for possible deflection.

Fig 34 The weight of a water tank must be distributed, and specially designed trusses are needed to bear this loading. Where water tanks are to be supported by trussed rafters the size, type and position of the tanks should be clearly indicated.

If tanks occur in the roof space, trusses must be specifically designed to carry the extra weight and the design would generally be based on spreading the load over three or more trusses with the loads being applied close to the node points.

Tank capacity	Member sizes (in mm) A & C	B	Max. trussed rafter span
Detail 'A'	50 × 75	2/38 × 100 or 1/50 × 125	6.50
300 litres max.	50 × 75	2/38 × 125 or 1/50 × 150	9.00
over 4 trusses	50 × 75	2/38 × 150	12.00
Detail 'B'	50 × 75	1/50 × 100	6.50
200 litres max.	50 × 75	2/38 × 100 or 1/50 × 125	9.00
over 3 trusses	50 × 75	2/38 × 125 or 1/50 × 150	12.00

Fig 35 Where headroom is limited, joist hangers are often used for the construction of a platform for supporting a water tank

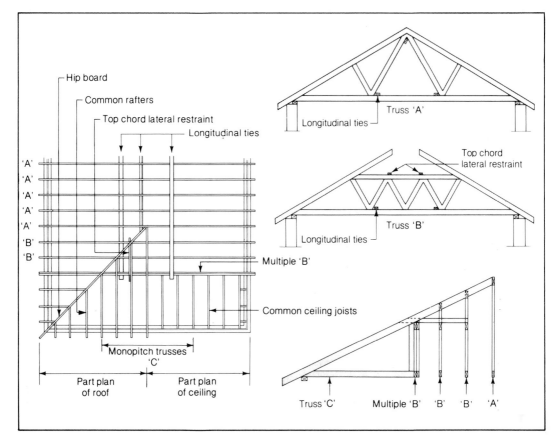

Fig 36 Guidance for erecting a hipped roof using Gang-Nail trusses. This Gang-Nail structural configuration has been designed to allow optimum economies in the fabrication of this particular type of roof:

Erection procedure
1 Erect standard trusses 'A' and temporary bracing up to the intersection of the ridge with the hip, following standard procedure. Occasionally on special designs the last standard truss may be a double ply in which case this should be nailed as described for 'T' roof junctions
2 Erect multiple ply girder in the correct position as indicated by the truss supplier
3 Erect intermediate hip trusses 'B' spaced at design centres and brace the horizontal top chords as required
4 Position hip board and trim truncated trusses to meet hip board snugly
5 Erect mono trusses 'C' and fix to wall plate and girder truss, trimming the extended top chord at the hip board. These trusses should be securely nailed to the bottom chord of the girder and to the top chord after the installation of a triangular section timber packing piece
6 Trim in the infil jack rafters ceiling joists
7 Complete roof following normal procedure

Fig 37 Diminishing valley jack rafter frames are used to produce a 'T' roof junction. Where there is no load-bearing wall through the intersection a girder truss will be required to carry the roof trusses over this opening. Due to the heavy loads being carried by these girders a larger than normal bearing is often required. It is recommended that consideration be given to the desirability of using a concrete padstone for the girder support. Gang-Nail Systems Limited or Gang-Nail Fabricators can supply information on minimum bearing areas and girder support loads for individual projects

Photo 37 Trussed rafters can be used in the construction of a hipped roof, although completing the operation is not an easy task for the amateur

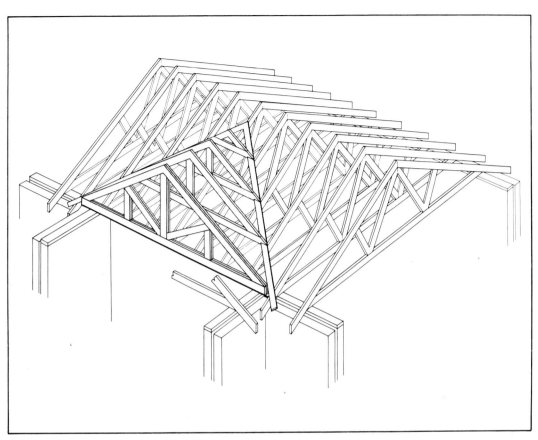

Hipped roofs

The design of units for hipped roofs varies from manufacturer to manufacturer. The layout usually involves the use of a flat-topped multiple girder and intermediate trusses; hip mono trusses are then used which are supported between this multiple girder and the end wall. A completed example is shown in the accompanying photograph, and more detailed construction procedures are given in manufacturers' data sheets. The erection guidance for trusses by Gang-Nail Ltd are reproduced in Fig 36.

T-roof junctions

Where roofs intersect, diminishing valley frames are used (*see* Fig 37). In the absence of a load-bearing wall, a girder truss is also required to support the trussed rafters which form the main roof. Construction detailing is provided in manufacturers' literature; for in-stance, procedures and diagrams are included in *Trussed Rafter Construction and Specification Guide* published by Gang-Nail Ltd.

Purlin and rafter roof construction

Before the advent of prefabricated roof trus-ses, roofs were constructed using a variety of configurations. For a more detailed descrip-tion of structures such as king post, queen post and mansard roofs, you should consult texts such as Alan Johnson's book *How to Restore and Improve Your Victorian House* (David and Charles, 1984). The concern here is with simple structures which a self-builder

Fig 38 A traditionally built duo pitch roof employs four principal components – a wall plate, rafters, purlin, and a ridge board. In large roofs, a collar brace may also be needed. A birdsmouth joint at purlins and wall plate will weaken the rafter if more than a third of the timber width is cut away.
Note: Dimensions given here are 'typical' rather than definitive. Variations will be found according to the design and size of the structure

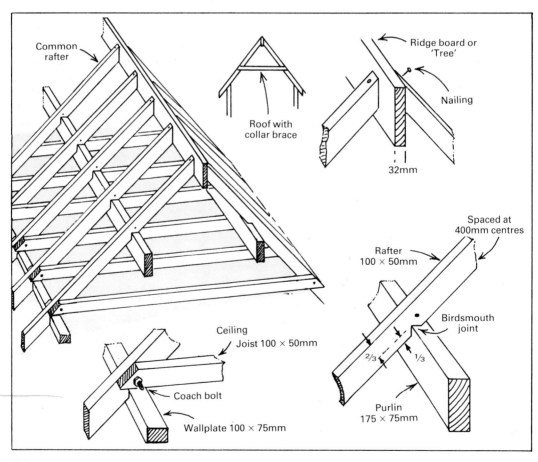

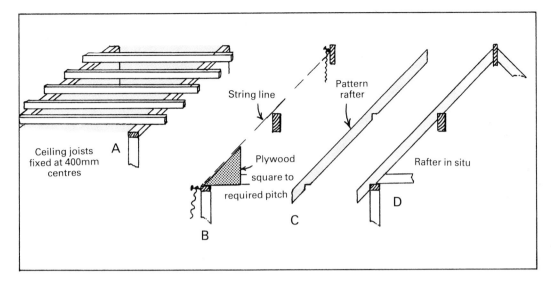

might tackle and for which detailed drawings have been prepared by an architect or building surveyor.

For a duo-pitch roof, the structure employs four principal components – wall plate, rafters, purlin and ridge board (or 'ridge tree'). Arguably this is an over-simplification, because in large roofs collar braces are necessary (*see* Fig 38). In old properties, struts can also be added to derive support from load-bearing partition walls – a contributor to structural integrity not often featured in modern houses. Reference is also made later (page 79) to the important contribution of ceiling joists which form a triangle in the configuration. As already mentioned, timber of special structural quality is required. This will be indicated on the drawings in terms such as MGS, SS, and MSS (*see* page 47). No less important is the size of the timbers needed to create the structure, and the designer will prescribe requirements on the drawings. When the Building Regulations 1976 were current, Schedule 6 in the appendix was the source for this data. The Schedule lists the performance of timber members according to size and provides information about purlins, common rafters, jack rafters and ceiling joists.

Constructing a roof is made considerably easier if the brickwork has been built accurately. Setting out the position of wall plates, purlins and the ridge board requires meticulous attention, and the amateur will not have the time to acquire techniques which the

Fig 39 Constructing a duo pitch roof can begin (A) by positioning ceiling joists on the wall plate at specified centre spacing. A string line (B) verifies the alignment of wall plate, purlin, and ridge board, whilst a plywood square can establish the correct pitch. A pattern rafter (C) is cut, and all subsequent rafters are shaped identically by using this as a template. Rafters placed in situ (D) should be set vertically with a plumb line or spirit level

tradesmen learnt from apprenticeship and experience. Traditional setting-out tools like the carpenters' square are complicated for the beginner to use, although detailed instructions accompany the Stanley product. A long steel tape measure is essential for establishing distances between plate, purlin and ridge board, and measurements taken across diagonals confirm if the configuration is square. You will also need a string line to verify the alignment of wall plate, purlin and ridge board as shown in Fig 39.

Rafters

If the ridge board, purlin and wall plate are placed in correct juxtaposition, identical rafters are required to bridge the spans. The procedure is to make up a pattern rafter against which all others are marked and cut (*see* Fig 39). At the purlin, a cut-out referred to as a birdsmouth joint is needed, and it is recommended that this does not reduce the width of the timber by more than a third (*see* Fig 38). A second birdsmouth joint is needed at the wall plate; you should leave ample eaves projection which will be trimmed later. At the ridge, the rafter must be cut at an angle to rest against the face of the ridge board.

The tradesman will undertake this work with speed and accomplished skill; the amateur must be prepared to take much longer. For instance, it may take several practice trials using offcuts of timber to establish the appropriate splay cut needed at the point of bearing between rafter and ridge board (*See* Fig 38). Cutting birdsmouth joints on the pattern piece to ensure that the rafter sits accurately and without bowing may take several attempts. Patience is rewarded because subsequent accuracy is dependent on this initial attention to detail.

Rafter fixings

Round wire nails are required to fix rafters in place. At intersections with both the purlin and wall plate, skew or cross nailing is the usual method of anchorage (*see* Photo 38).

Photo 38 At the birdsmouth joint at the purlin, rafters are held in position by 'cross-nailing'. Wire nails driven at an angle from both sides of the rafter provide adequate anchorage

Purlin

The function of a purlin is to carry the weight of all the common rafters and their covering material. By acting as an intermediate support point, it effectively reduces the distances that a rafter is required to span. Of necessity it must be sturdy timber, and may need plenty of muscular assistance to lift into position. The ends should be treated with a proprietary preserver such as Cuprinol, and before the purlin is finally bricked into place, the plate, purlin and ridge measurements must be accurately finalised.

Ridge board

The ridge board represents a bearing point for rafters, and in view of its role is usually a timber of modest size. However, if it is to remain straight when 'eyed' along its length, rafters must counterbalance and be of identical lengths. With rafters of suitable girth, this feature gives a self-supporting structure.

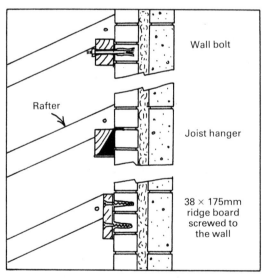

Fig 40 Alternative methods of supporting the upper end of rafters on a lean-to roof which is being constructed traditionally

Lean-to construction

The porch area shown in the photograph of the self-build house (*see* Photo 35, page 63) comprises a 'lean-to' construction. At their upper end, rafters are supported on a wall plate which is fixed to its vertical supporting

78

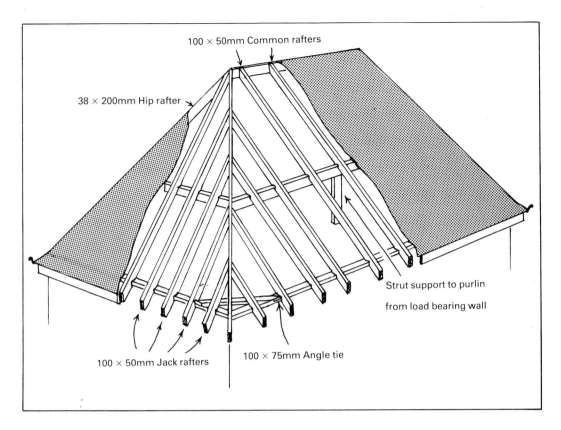

100 × 50mm Common rafters

38 × 200mm Hip rafter

Strut support to purlin

from load bearing wall

100 × 50mm Jack rafters

100 × 75mm Angle tie

wall with expanding wall bolts (*see* Fig 40). Alternative methods are to use a ridge board or purpose-made metal support hangers. The rafter is usually notched with a birdsmouth joint to sit on the plate.

Ceiling joists

Whereas rafters press together at the ridge board to give a self-supporting structure, this arrangement which is described as a 'couple roof' tends to force the support walls outwards. To the builders of medieval cathedrals, this was a constant problem. The Normans solved this by sheer bulk in the supporting walls, but in their quest for refinement the later Gothic builders found answers using buttresses – whose functional role was accompanied by ornate treatment. Neither answer is suitable in our domestic context, but the addition of ceiling joists, bolted across the base of the rafters gives the solution. A series of triangular configurations formed by rafters and ceiling joists yields what is referred to as a closed (or 'close') couple roof. Suffice it to say that ceiling joists are particularly important in providing structural integ-

Fig 41 Principal structural timbers which form a hipped roof built traditionally. Modern construction usually favours the use of pre-formed trusses, but neither task is easy for the amateur self builder

rity, and over large spans the designer might specify sturdy coach bolts (eg 10mm/³⁄₈in) to bolt up joists with rafters. On short spans, nails are deemed adequate, though it is often recommended to pre-drill with slightly undersize holes.

Hipped roofs

It is unlikely that an amateur would tackle the construction of a traditionally built hipped roof, although some indomitable self-builders have tried, and succeeded. The construction is more complex, and poses several more difficult tasks for the carpenter. As shown in Fig 41, hip rafters spring from the corners of the building to the ridge board. Bearing against these members are jack rafters whose lengths are dissimilar. One of the more difficult tasks for the carpenter is cutting the upper end of jack rafters where they intersect with the hip rafter. The sloping cut is a composite one which necessitates both

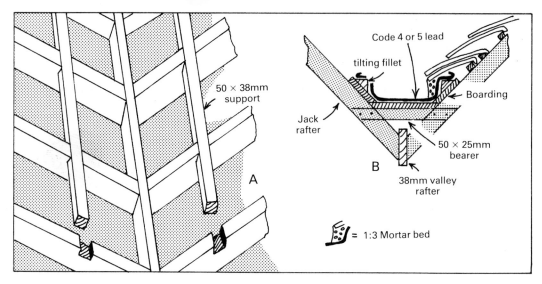

Fig 42 Preparatory groundwork for an open valley gutter. (A) shows a framework which will support sections of pre-formed plastic valley gutter. The groundwork in (B) shows the continuous support boarding needed for a lead-lined valley

skew and splay cutting to bear against the hip rafter with full alignment. Additional bracing is also needed at the corners because a hipped roof places stress on the quoins (corners) of the supporting walls.

Valleys

The point where two roofs intersect is called a valley; in several respects, it resembles a hip construction which has been inverted. Hence it features a longitudinal member, called the valley rafter, which creates the fall line down the centre of the valley. Bearing against this central timber are jack rafters which again feature skew and splay trimming seen on hipped roofs. Preparatory structure varies according to the type of valley covering intended – lead, plastic valley gutter, purpose-made valley tiles and so on. The accompanying illustration (Fig 42) shows the groundwork needed to support a uPVC valley gutter, for which supplementary 50 × 38mm (2 × 1½in) timbers have been let into the jack rafters on either side of the central valley timber. Timber valley boards may be required to provide a support if trough valley tiles are used, and tiling manufacturers show detailing of preparatory groundwork in their technical literature. Similar support is

needed for lead-lined open valleys (*see* Fig 42), although an additional angle fillet is needed on either side to channel the rainwater.

Finishing both trussed-rafter and traditional roofs
Negotiating chimney stacks

If a chimney stack is built in a position which will require it to pass through a roof, a back gutter must be constructed and the structure must be trimmed to form an opening. Section

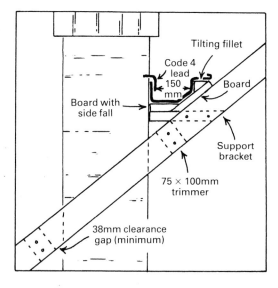

Fig 43 Timbers known as trimmers must be positioned where a chimney stack passes through a roof. Preparatory groundwork must also include the construction of a shaped back gutter which will provide support to lead flashings

L10 of the Building Regulations 1976 stated that the gap between the stack and trimmers should be no less than 38mm (1½in) (*see* Fig 43). Specification based on the post-1985 regulations will be given in the prepared drawings, although in a trussed rafter roof a chimney stack is often sited to pass *between* rafter spacings. If this positioning is impossible, details showing how to construct an opening with the Gang-Nail trussed system, reproduced from *Trussed Rafter Construction and Specification Guide,* is given in Fig 44.

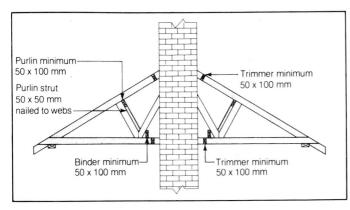

Purlin minimum 50 x 100 mm

Purlin strut 50 x 50 mm nailed to webs

Trimmer minimum 50 x 100 mm

Binder minimum 50 x 100 mm

Trimmer minimum 50 x 100 mm

Fig 44 Special provision has to be made to negotiate a chimney stack. Whenever possible, chimney openings should be accommodated within the trussed rafter design spacing. When this is not possible the method illustrated below should be used

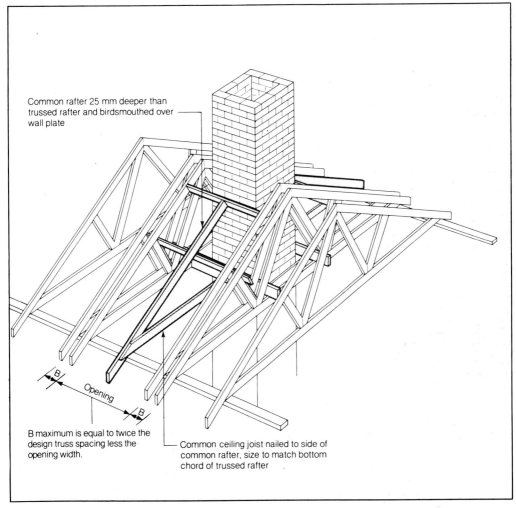

Common rafter 25 mm deeper than trussed rafter and birdsmouthed over wall plate

B

Opening

B

B maximum is equal to twice the design truss spacing less the opening width.

Common ceiling joist nailed to side of common rafter, size to match bottom chord of trussed rafter

Gable details

In terrace housing, a roof may terminate at a party wall which projects above the covering tiles or slates to form an abutment. However, where a roof covers a gable end, several alternative treatments are possible. The cheapest and least attractive finish is to terminate the roof in line with the face brickwork of the gable wall. Absence of timber barge boards eliminates maintenance, but the lack of overhang means that the roof offers no weather protection to the walls. An answer which is both maintenance-free and functional is to corbel the brickwork along the peak of the gable. As shown in Photo 39, the idea of stepping the brickwork outwards (referred to as corbelling) can give a long-lasting and attractive finish.

Notwithstanding the paint-saving benefits of corbelling, projecting verges in timber are more fashionable. Construction is not difficult, but you need to work in close co-operation with the bricklayer. Carrying the structure from the last rafter or truss through the gable end wall to provide supports for barge boards requires short cantilever rafters. This structure is sometimes called a gable ladder on account of its appearance. Its rungs (the cantilever rafters) are nailed to the last common rafter at approximately 450mm (18in) centres, and 100 × 50mm (4 × 2in) timber is often used. In order to seal off the roof void, the cantilever rafters must be enclosed in masonry. As shown in Photo 40a, a string line provides guidance so that bricks can be marked off, cut (*see* Photo 40b), and placed at the correct pitch. Throughout this stage, you must ensure that the short rafters are held in alignment.

When the exact verge overhang has been decided, the cantilever rafters must be trim-

Photo 39 In spite of the popularity of timber barge board, corbelled brickwork represents an attractive and maintenance-free alternative at the verge of a roof. This example, completed in 1906, still looks attractive many years after completion

Photo 40a By marking off against a string line, the brickwork at this gable is cut to match the finished pitch of the roof

Photo 40b In skilled hands, a club hammer and bolster is sufficient to make sharp cuts. Beginners might find more success with a tipped handsaw designed to cut bricks

C ◄

A ◄   ◄ B

Photo 41a To construct a projecting verge, short cantilever rafters must be taken through the gable. Brickwork will be built around them to provide support, with a regular monitoring of their alignment

Photo 41b Cantilever rafters must be cut to identical length using a stretched cord line to mark the sawing point. An 'outer rafter' (arrowed) is then nailed into place

Photo 41c Using the support of the sturdy outer rafter, a timber barge board is nailed into place. This must be wider than the outer rafter to provide enclosure for a gable soffit

med to an exact length. This is done by precise marking using a string line stretched tightly across the top of *all* projections. Having sawn off the surplus, an outer rafter – usually in 100 × 50mm (4 × 2in) sawn timber – is nailed to the cantilever rafters, taking care not to disturb the gable brickwork. Some builders omit an outer rafter, and fix the final barge board direct to the projections. This is bad practice, because the timber – usually 25mm (1in) nominal – is inclined to flex as it bridges the supports. Several materials can be used for soffits, but external-grade 9mm (3⁄8in) plywood is easiest to fit. This is nailed to the underside of cantilever and outer rafters *before* the barge board is added. This allows it to be planed flush with the outer rafter (*see* Photos 41a, 41b and 41c). Few of today's builders have time to create ornamental barge boards, but I have seen the work of one self-build amateur who decided to produce attractive cut-outs using a router. If you have some spare time, a penchant for woodwork and the appropriate machinery, there is no reason why such decorative refinements should not be revived.

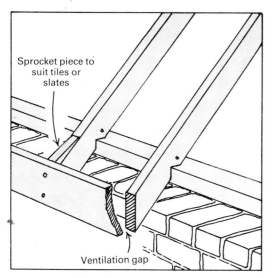

Sprocket piece to suit tiles or slates

Ventilation gap

Fig 45 Eaves finish may feature flush fascia boards. This reduces maintenance, but also lowers weather protection of the face brickwork. In modern practice, where a ceiling level insulant has been installed, a ventilation gap is important

Photo 42 Fascia and soffits need constant painting or preserving, whereas decorative corbelled brickwork offers an attractive maintenance-free alternative at the eaves. The preparatory structural work is shown here in detail

Eaves detailing

Like gable finishes, a variety of treatments are possible at the eaves. Recent concern for ventilation in roof spaces has influenced design, and the inclusion of grilled eaves vents is discussed in Chapter 7. In order to reduce both installation and maintenance costs, some houses are built with no eaves overhang (*see* Fig 45). A fascia board may be nailed to the wall to offer a mounting for gutter brackets, but even this can be omitted if spiked gutter brackets are driven into the mortar. There is some saving from this finish, but aesthetically the lack of an overhang never looks comfortable. Moreover the greater exposure to weather is a feature of disadvantage, both for the walls as well as for the home owner.

An overhang at the eaves is not a difficult feature for the self-builder to construct (*see* Fig 46). Firstly you must mark off the rafters to the required length using a string line as shown in Photos 32a-32c, page 53. A spirit level held vertically gives the marking line for a splay cut, and careful sawing produces a fixing point for the fascia board. If you

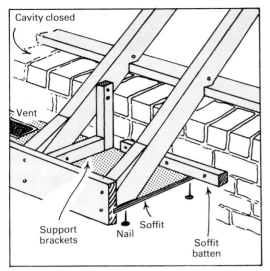

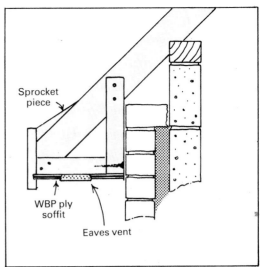

Fig 46 Constructing overhanging box eaves can be done in several ways. Soffit in WBP ply is popular and can be fixed at one end in a fascia board groove. At the other, support brackets give a sound construction, but a wall fixed batten is another means of attachment

Fig 47 Ventilation at the eaves can be achieved using various proprietary types of soffit grills. Elevation of the fascia board, and infill tilted sprocket pieces are required to lift the bottom course of tiles or slates. The manufacturer of the cover units will give required dimensions

install a horizontal soffit, its position depends on the pitch of the roof and the length of the rafter extension. The former is already established, but by shortening the rafter, soffit height can be increased. However, this is accompanied by a corresponding decrease in its width.

Fascia boards

These are usually made from planed softwood, and joins are made with a sloping face known as a scarf joint. If shrinkage occurs, the conspicuous gap which would appear between a butt joint is thus disguised. Fascia boards must extend above the top of rafters and be completed with a sprocket piece in accordance with the requirement of the type of tile or slate used (*see* Fig 47). This is to support the lowest eaves tile/slate (*see* page 110); the projection will be somewhere between 30mm (1¼in) and 69mm (2¾in), depending on pitch angle and the type of tile. Information is given in manufacturers' literature; for example, Marley Roof Tiles give a table listing requirements for their different coverings according to pitch.

The thickness of fascias is generally 25mm (1in) (nominal), and if a ply or fibre-cement soffit is fitted, the fascia is usually grooved to

provide a housing for the soffit board. Grooved fascias can be supplied to order by a good timber merchant, or you can cut a slot by passing a portable circular saw across the board several times. However, a number of proprietary forms of ventilator strips are now made to attach to the rear of fascias, and if you adopt one of these ventilator systems a housing slot for soffit board is usually incorporated in the design of the unit.

Soffits

On occasions, it is decided to leave the undersides of eaves exposed. The classic example is shown in the eaves of Swiss chalets. However, in this country fitting a soffit is a more common strategy in modern building. Soffits used to be made from tongued and grooved boarding which involves rather more work to install, and the tongues are liable to dislocate if the timber shrinks. Much easier to use is exterior-grade 9mm (⅜in) ply; fibre-cement boards are sometimes preferred, although they damage easily during fixing.

Whereas a soffit board is fixed to the fascia in a groove, or within strip ventilator sections, it is generally held at the wall by nailing into a 32 × 25mm (1¼ × 1in) soffit batten (*see* Fig 46). You need to fix the batten at the appropriate height using either masonry nails or screws, and wall plugs. Alternatively it can be fixed by strutting up to the rafters with short battens used as support brackets.

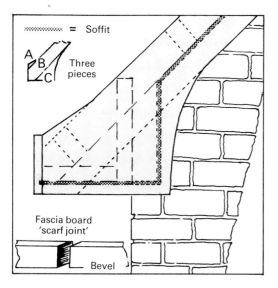

Fig 48 Where a boxed eaves construction meets a gable corner, additional sections are fixed to the barge board to provide a total enclosure. Note the line taken by the soffit in transition between eaves and verge

Intersection of fascia and barge board

As shown in Fig 48, a sprocket piece (A) is needed to give continuity at the corners of the building so that the projection of the fascia board aligns with the upper edge of the barge board. An additional section (C) may also be needed on the underside of the barge board. Working out the detailing to provide an enclosure at the junction of gable and eaves units is generally completed on site.

Timber preservation and alternative finishes

Wood staining has become a popular alternative to painting, but whichever form of preservation is chosen a good builder will ensure that the rear surfaces of fascias and barge boards are also treated. Fascia fixing should be done with galvanised nails which are driven below the surface with a punch. The holes are then filled with an exterior-grade compound to suit the chosen finish.

Unfortunately, the maintenance of eaves and gables involves time, money and paint. As the quest to reduce maintenance work continues, it is appropriate to mention that several alternative finishes can be used in preference to timber constructions. For example, Marley has introduced a uPVC fascia, barge board and soffit system with integral ventilation. Few tools are required to construct the eaves and gable finishes, and the product complements Marley's uPVC dry-fix verge and ridge system. Following on from a heavy involvement in the manufacture of uPVC rainwater systems, this manufacturer is now pursuing developments in plastics at a special division, Marley Extrusions Ltd. There will undoubtedly be an increase in the use of maintenance-free uPVC in building, and for further information on this roofing product the address of Marley Extrusions is given in Appendix 2.

Another development which gives an alternative finish to gables is the Redland cloaked verge tile, described on page 129. It is shown in Photo 76a that this is a pleasant maintenance-free alternative to barge boards. A similar cloaked finish is often seen in Wales in which the gable enclosures are made with slabs of slate. However, it is pointed out in Chapter 6 and shown in Photo 116 that this finish is not always completely weatherproof.

Summary

Of necessity, it is only possible to give a brief survey of timber structures on pitched roofs. Sources for further guidance have been given, but if you need more detailed information on carpentry and joinery you should consult the texts written for trainee architects or students of building.

5 Pitched Roofs – Coverings Compared

Thatch, slates, clay tiles, concrete tiles, profiled tiles, plain tiles and pantiles are just a few examples of coverings used for pitched roofs. Each has its attributes and failings, and this chapter compares features such as appearance, functional performance and methods of installation. In the event of a self-build enthusiast contemplating roofing work – even if it is only a porch, bay window, or garage – it is important to realise that pitch, roof shape, degree of exposure and a number of other constraints limit the choice of covering quite significantly.

This chapter is therefore concerned with theory rather than practice. Chapter 6 puts this knowledge to use, and gives detailed guidance on how to tackle a roofing job using either tiles or slates. It explains the traditional method of installation, and also looks at the modern 'dry-fix' techniques.

Roof-covering terminology

The accompanying diagrams (Figs 49-51) explain many technical terms, but these should be noted in conjunction with the following points. Words already included in Chapter 1 and relating to roof shapes are omitted; re-check Fig 1 (page 17), for example, if you are uncertain of the difference between a hipped roof and a duo-pitch roof.

General terms

Left- and **right-hand verges** Their designation is given when looking directly at the roof slope from ground level.
Eaves tiles These are singled out for recognition because they are the first units to be

Fig 49 General terms used in roofing

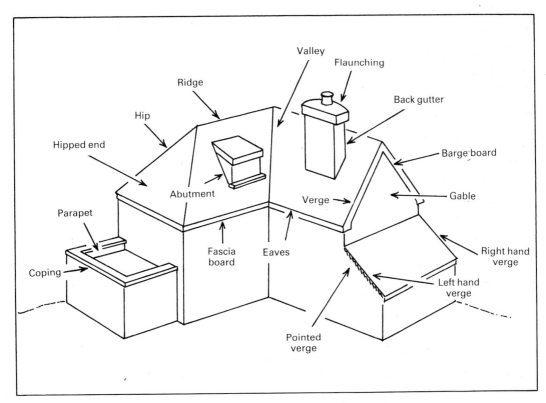

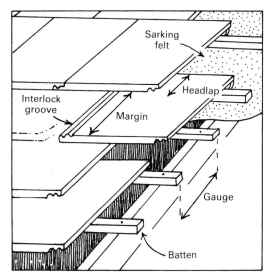

Fig 50 Terminology associated with a single lap roof covering, eg tiles with a moulded interlock groove

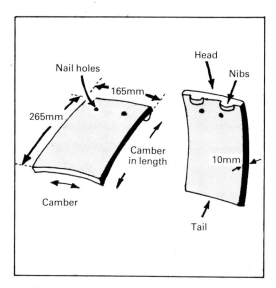

Fig 51 The anatomy of a plain tile

laid; similarly in thatching, the first bundles of reed are fixed at the eaves. In double-lapped roofs, there will be a double or treble course of tiles at the eaves.

Abutment   Often the covering material will be bounded by a vertical edge. This may be a parapet, the cheeks of a dormer window or the sides of a chimney stack.

Double barriers and groundwork

In modern building, pitched roofs are given two separate coverings – thatch excepted.

Below the outer material, the installation of a secondary barrier or underlay is mandatory. Often the secondary barrier is a tearproof reinforced bitumen roofing felt, designed specially for the purpose. This option is sometimes referred to as 'sarking felt' or 'slating felt', and the next chapter describes underlays in greater detail. Both coverings need a secure base, and materials fixed to the supporting structure are referred to as the 'groundwork' or 'substrate'.

Laps

An outer covering made up of units, for example slates or tiles, calls for an overlap of each adjoining unit, referred to as 'lap' (see Fig 50). Lap on the upper and lower portions is called 'head lap', which can be increased or decreased by altering the spacing of the battens on which the tiles or slates are fixed. This spacing is called 'gauge', and on account of head lap, only part of each tile or slate is visible on the surface. The exposed portion of tiles and slates is the 'margin', and the vertical measurement equals the gauge of the battens.

To be fully waterproof, there must also be 'side lap', and on moulded tiles this sometimes takes the form of an interlocking groove.

Slate and plain-tile anatomy

Whereas Welsh slate is flat, the rectangular 'plain tile' is moulded with gentle cambers. Longitudinal and cross-camber is to prevent 'rainwater creep' – the capillary action in which rainwater is drawn upwards and sideways between the lap of adjacent units.

As shown in Fig 51, 'heads' and 'tails' refer to the upper and lower ends of slates or tiles; 'nibs' are found on tiles only. When looking at Fig 52, notice how plain tiles and slates have a 'double head lap', which means the thickness of three tiles is achieved at the head.

Profiled-tile anatomy

Shaped tiles, referred to as 'profiled tiles', are based on traditional styles used in Mediterranean countries (see Fig 53). The

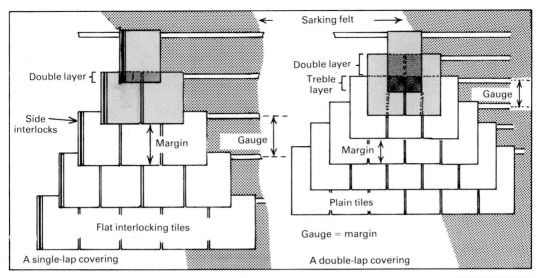

Flat interlocking tiles

A single-lap covering

Plain tiles

Gauge = margin

A double-lap covering

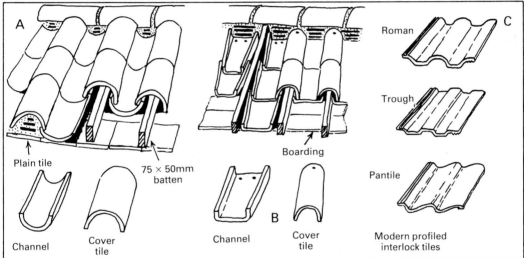

Fig 52 Roofing with interlocking tiles requires fewer units and is known as a single-lap covering. A double-lap covering, eg slates of plain tiles gives a coverage of at least double thickness; but many more units and battens are required

Fig 53 Traditional tiles used in Mediterranean countries are shown here – (A) Spanish style, and (B) Roman style. Modern versions (C), which adopt similar profiles, feature an interlock channel; they are classified as single-lap tiles

modern types have locating channels at the sides, which produce a weatherproofing overlap called a 'side interlock'. This form of construction is not limited to profiled tiles; flat tiles can now be obtained with side interlocks as well.

Small tolerances are built in to the side lap, and the opportunity to squeeze adjacent tiles closer together, or to pull them apart without disengagement, is called 'sidelock shunt'. Typically, this only amounts to 2–3mm (about ⅛in per tile), but accumulatively over a full course, this adds up appreciably and provides valuable room for adjustment.

Traditional terms

Before the use of underlays, the risk of driven snow or rain being forced underneath tiles or slates was reduced by 'torching'. The term referred to the application of mortar on the underside of the covering units, but this precautionary measure is not needed with modern techniques.

You may also hear reference made to 'peg

Photo 43 Pegged stone slates laid at random and in
diminishing courses on the large medieval roof of
Stokesay Castle, Shropshire

tiles' or 'peggies'. Before the arrival of nibbed units, plain tiles were hooked over battens by means of oak pegs fastened through holes in the units. This technique is shown in the accompanying photographs (*see* Photos 43 and 44) taken at Stokesay, Shropshire, and testifies to the durability of oak. Another variation was to use sheep's teeth, which speaks rather highly for the durability of their dentures.

The function of the covering material

The material used to cover a roof represents the first line of defence against weather, which in the British Isles is a tall order! All kinds of contingencies are considered in roof design, and BS 5534 (Part 1 : 1978) requires that the cladding and its associated fixings could cope with the worst wind loadings that might be expected within a period of fifty years. Wind force involves both pressure and suction, and if a wind blows across a roof there will be a suction effect on the lee (shadow) side, and pressure on the windward side. The covering must also contend with the sharp impact forces of hailstones – which in freak storms may be as large as walnuts. Driving rain and blizzards pose a further threat, and as temperatures rise and fall, the covering must have the ability to cope with expansion and contraction. If the covering is too absorbant, the ingress of precipitation moisture will hasten destruction, especially when frost causes expansion within the material. Although the need for a sound integrity is unquestioned, it isn't always appreciated that 'the elements' encompass so many multifarious forces of destruction.

Another function of the covering material is its role as a barrier against fire (*see* page 14). Resistance to solar radiation and ultraviolet rays is another requirement, and this is linked with heat gain in the roof space which can pose problems. Moreover, a roof laid in modern times must also be able to resist atmospheric pollution. Finally, a problem as old as time concerns the destructive powers of vegetation, birds, insects and rodents which represent another challenge to the roof over our heads. To produce satisfactory coverings, manufacturers have to fulfil many requirements, and will subject their products

Photo 44 Oak pegs which hold the stone slates in place on Stokesay Castle, Shropshire, have endured the test of time

Photo 45 Adorning a building dated 1615, 'Horsham Stone', quarried many years ago in Sussex, is a traditional sandstone 'slate' of bold character

to rigorous tests such as wind-tunnel experiments. However, their efforts will be in vain if you fail to install the covering in accordance with good practice.

The appearance of the covering material

Functional integrity is not the only matter of concern; appearance is also important. It is appropriate to preserve regional character whenever possible, and local authority planners often place restrictions on the choice of covering material. Many parts of Britain have distinct roofing styles, and the examples shown here of Horsham stone in Sussex (Photo 45) are pleasantly different from the fine coloured sandstones used in parts of West Yorkshire (Photo 46). Regional character can sometimes be preserved using reclaimed materials, but where this proves difficult a number of modern imitations are pleasingly convincing. The photograph of a roof covered in Bradstone reconstructed Cotswold stone illustrates this well (Photo 47).

It is important to add, however, that certain coverings look incongruous when used away from their *genius loci*. For example, thatch is essentially a country material and whereas the accompanying photograph (Photo 7, page 15) shows fine workmanship, this thatched property is rather misplaced in the urban setting of Leighton Buzzard. And no one would argue that an unusual property roofed in seaweed would be a veritable 'fish out of water' if moved from its location on Pembrokeshire's south coast.

Determinants of choice

Geographical location

Weather variation, coupled with topographical differences, is sufficiently marked in Britain to influence the choice of covering materials. The degree of exposure must be

Photo 46 This urban setting in Rothwell, near Leeds, shows large, colourful stone slates laid in random widths – a typical covering in this part of Yorkshire

Photo 47 This modern reconstructed stone is a convincing copy of traditional Cotswold slate

taken into account, and a seaside bungalow poised high on a Cornish cliff will receive the full force of prevailing winds, and considerably more rainfall than a similar property in a London suburb. With the help of the Meteorological Office and the Irish Meteorological Service, manufacturers of roof coverings collate information about areas which suffer from adverse weather, and the data is analysed and meticulously documented. In their specification data, matters such as minimum-pitch requirements are often postscripted with alternative data for exposed sites. In practical terms, this may mean that the frequency of clipping or nailing roof tiles will be increased – a requirement clearly stated in product literature and noted by an architect when designing a property.

Roof shape

A roof which features tight curves cannot be covered with large section tiles like interlocking pantiles. If you look at a roof with an eyebrow dormer like the one illustrated in Photo 3 on page 11, it will probably be covered with small plain tiles. Similarly, a property with multiple pitches, tiled dormers and intricate detailing calls for a great deal of cutting and shaping if large tiles are used. In particular, the curved profiles of modern pantiles or Roman tiles are very difficult to cut and fit around intricate detailing. Plain tiles are much easier to use, and small slates are particularly suitable if you tackle a roof with complicated shapes.

Support structure

Coverings are dissimilar in weight, and in any reroofing project it is always necessary for the supporting structure to be checked by a building surveyor, particularly if refurbishment involves the use of a heavier covering. For instance, double-lap coverings (eg plain tiles) impose more weight than a single-lap arrangement. Moreover, concrete tiles are generally heavier than their clay counterparts. It is the light weight of synthetic slates which has contributed to their recent popularity, although this is offset by their life span which is short compared to natural slate. Table 1 (Appendix 1) gives a rough indica-

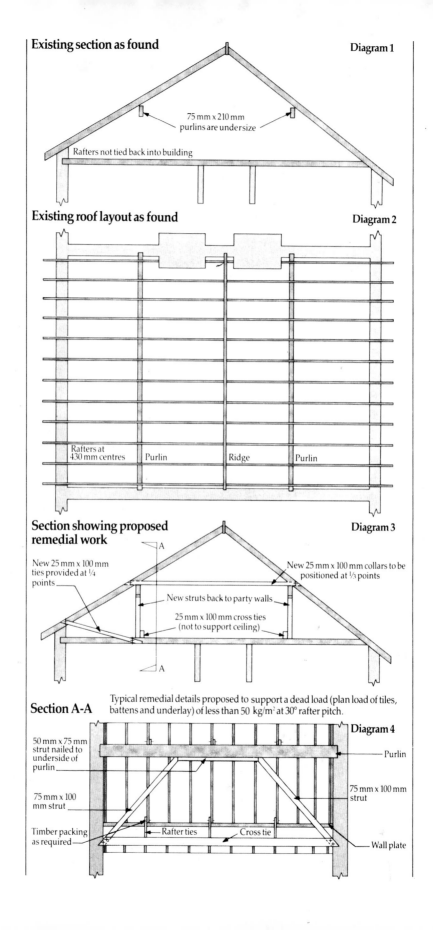

Existing section as found   Diagram 1

75 mm x 210 mm
purlins are undersize

Rafters not tied back into building

Existing roof layout as found   Diagram 2

Rafters at
430 mm centres   Purlin   Ridge   Purlin

Section showing proposed   Diagram 3
remedial work

A

New 25 mm x 100 mm
ties provided at ¼
points

New 25 mm x 100 mm collars to be
positioned at ⅓ points

New struts back to party walls

25 mm x 100 mm cross ties
(not to support ceiling)

A

Section A-A   Typical remedial details proposed to support a dead load (plan load of tiles,
battens and underlay) of less than 50 kg/m² at 30° rafter pitch.

Diagram 4

50 mm x 75 mm
strut nailed to
underside of
purlin

Purlin

75 mm x 100 mm
strut

75 mm x 100
mm strut

Timber packing
as required

Rafter ties   Cross tie

Wall plate

94

Fig 54 Roof structures vary considerably. The details shown on this page are derived from terraced houses surveyed in the Midlands and show some typical problems found, with proposed solutions.

1 and 2 show the roof as found. 3 and 4 illustrate the same roof after strengthening work has been carried out.

Intermediate support is provided to the purlins by means of diagonal struts, with the compression loads transferred to the brick party walls at each end. This reduces the effective span of the purlins and thus the bending stresses.

Buckling stresses in the purlins are reduced by providing lateral restraint in the form of collars, placed at intervals across pairs of common rafters.

Shore ties are fixed at intervals from ceiling joists to adjacent rafters, to tie the roof structure into the external walls. Packings are provided as necessary where joist and rafter are not aligned.

Binders, supported at intervals from the purlins, are fixed across the ceiling joists in the same vertical plane as the purlins, and tied into the brickwork at each end to provide bracing and additional support at ceiling level.

Roof strengthening, such as shown in these details, would be required to meet current Building Regulations for any roof coverings with a dead load on slope (after the imposed load is taken into account) of up to 50kg/m². Heavier roof coverings could require additional strengthening.

Rafter sizes should be checked with schedule 6 of Building Regulations, but are often found to be adequate. Remedial work is not normally required unless inadequate purlins have caused permanent distortion

tion of weights per square metre (10¾sq ft), although you should check exact details relating to specific products. Moreover, alterations in lap will affect the final weight factor. In refurbishment work, 'firming up' the rafters or purlins is often necessary, and although this isn't usually a difficult task, it underlines why expert advice is needed. Manufacturers will also advise, and an example of a typical strengthening procedure is shown in the free booklet *Grant-aided Housing Rehabilitation* published by Redland Roof Tiles (*see* Fig 54).

Roof pitch

Roof pitch may also exclude some covering materials. For example, hand-made clay plain tiles should not be laid at pitches less than 45 degrees. With their tendency to absorb rainwater more readily than a machine-pressed equivalent, their use on a shallow-pitched roof discourages rapid water dispersal; if moisture is absorbed, this increases the risk of frost damage. Machine-made concrete plain tiles on the other hand

can usually be laid on pitches as shallow as 35 degrees. When considering profiled concrete interlocking tiles, their performance is influenced by the nature of their surface finish. For example, a typical Roman profile tile with a textured granular face might be limited to roofs no flatter than 30 degrees, whereas its identical counterpart with a smooth surface would perform acceptably at pitches down to a minimum of 17.5 degrees. Also suitable on pitches as low as this is a flat version of the interlocking tile, and the combination of a smooth and level surface is particularly conducive to fast run-off in regions of high rainfall levels.

So far, the references to pitch have focused on the minima for different types of tile. With regard to steep roofs, the profiled tile is less versatile and is usually only suitable for pitches up to 44 degrees, though this can be increased with special fixing. In contrast, the plain tile sits happily on pitches around 60 degrees, and if every tile is nailed there is no upper limit. Hence it is used in a completely vertical mode as cladding on exterior walls, or on the side cheeks of a dormer window.

Nailing slates

The performance of slates is determined by their size, thickness and whether they are head or centre nailed. For instance, small slates measuring 305 × 205mm (12 × 8in) are considered to be unsuitable on pitches less than 45 degrees – especially the thicker Lake District slates, which do not bed down closely. However, a pitch as flat as 22 degrees can be covered successfully using large 610 × 355mm (24 × 14in) Welsh slates fixed with centre nails. Like plain tiles, slates can also be used vertically as wall cladding, hence there is no upper pitch limit.

As a caveat, however, it is important to add that the pitch limits quoted above assume a sheltered site, and performance data should always be checked in manufacturers' literature for specific products.

Cost

This is a significant factor, and costs vary considerably from covering to covering. Comparisons are usually worked out per square metre (10¾ sq ft) of actual roof area – which is larger than the plan area shown on the drawings. You should exercise caution when comparing data in advertising literature; check whether the figure includes labour costs, or the inclusion of underlay, battens and nails. Annual price guides are published for the building industry which you could obtain from the public library; another source is the biennial publication *Specification – Building Methods and Products* (Architectural Press).

In brief, a troughed single-lap concrete tile is one of the least expensive options. Synthetic slates are around fifty per cent more expensive, whereas new Welsh blue slates may be three times the price of the least expensive concrete tile. One of the most expensive coverings is the hand-made clay tile, which may work out at four times the cost of an interlocking concrete tile. In part, this is due to the extra quantity used in double lapping, but in certain settings the appearance of this particular roofing material is noteworthy. Cost is not the only factor to consider in a comparison.

Covering materials compared

Whether you contemplate self-build work, a refurbishment operation, or calling on the services of a roofing contractor, it is useful to be aware of the strengths and weaknesses of different covering materials. Some of the more common examples are critically evaluated in this section.

Thatch

Thatch is a term used to describe the many humbler materials used for covering roofs – including heather, brushwood, reeds, straw and other natural vegetation. However, this appraisal is confined to reed or straw roof-covering materials which are most common. Since the recent revival of interest in country crafts, thatch has received considerable attention. Certainly in the present age of energy conservation, its good performance as an insulant cannot pass unnoticed. But at the risk of being controversial, it must be stated bluntly that thatch is functionally obsolete, and fulfils no useful purpose which

cannot be better achieved using tiles, sarking felt and a modern loft insulant.

Thatch suffers from many disadvantages. It requires materials of appropriate quality that are difficult to obtain in quantity. The best material is reed, and the type which grows in the shallower waters of the Norfolk Broads is particularly suitable. This is cut after the leaves have been killed by frost, and is then harvested in January or February when the weather is at its least pleasant.

In application, thatch is only suitable on roofs with a steep pitch – 50–55 degrees is preferred – and its installation cannot be made more economical using industrial methods. Rainwater is shed from the eaves to the ground below, and it is very rare to find any provision of guttering. With regard to life span, straw thatch will last some ten to twenty years, combed wheat straw about thirty to forty years, whilst even Norfolk reed will probably require total replacement after fifty years. Unfortunately, however, the ridge needs attention rather more frequently. Unlike the main body of the roof, this requires more pliable material such as sedge or rushes, which are held down with lengths of hazel or willow. The method of fixing is usually left exposed, and the wooden staples are woven in attractive patterns. This part of the roof usually needs replacement after twenty to twenty-five years, but repairs can be done without upsetting the reed or straw beneath it, as shown in Photo 48.

Among its various disadvantages, thatch is prone to infestation, and offers a splendid abode for nesting birds or home-seeking rodents. On account of fire risk, it was forbidden in London as early as 1212, banned from Hull in the reign of Elizabeth I, and disallowed in Cambridge by an Order of the Council in 1619. Notwithstanding this legislation, it has been argued that risks are exaggerated and insurance unjustifiably inflated. Only where buildings are in close juxtaposition is the risk heightened. Fire retardants can be used but are not generally approved because it is thought they hasten the onset of mould growth. Moreover, chemical retardants can easily be washed away by heavy rain. A more certain provision is a sprinkler system which is installed on buildings of historical importance.

When noting disadvantages, it will be appreciated that thatch owes its survival to just one feature – its appearance. Ironically it has not always been admired and was once regarded as a covering for properties whose owners could not afford tiles. In response to this view, less affluent home owners would tile the pitches facing the road but 'make do' with thatch at the rear. Attitudes change, and today there are many who feel that no other covering can match the elegance of its soft, flowing lines, or stimulate such rich images of rural landscapes.

With regard to thatching techniques, one must remember that this is a primitive – though skilful – tradition pre-dating printing and the widespread dissemination of information. In consequence, thatching techniques differed from locality to locality just as much as materials and the roofers' vocabulary. Subtle differences reflect regional style and the individuality of the craftsman. In general terms, the procedure is

Photo 48 A new ridge on an older covering. Ridges on thatched buildings require more frequent replacement, and require a pliable material like sedge or rushes which flex to the shape of the peak

to secure tight bundles of thatch to the roof, starting at the eaves. There are four main methods for such a fixing: bundles may be sewn with twine; they may be held with staples and rods lying across the roof; they may be thrust into or between turves; or, as a more primitive method, held down with ropes across the surface. It is important to achieve a dense thatch for the covering to be weatherproof and, whichever method of fixing is used, it must ensure sound anchorage.

There *are* amateurs who have had successful attempts at thatching, but the problem usually reported is difficulty in obtaining materials. For most people, the best strategy is to seek the help of experts whose life is their craft. Advice can be obtained from a number of sources, and the address of the

97

Thatching Advisory Service is given in Appendix 2. Recently the Consumers' Association also drew up a register of the secretaries of local Master Thatchers' Associations. Alternatively you can obtain advice at your County or District Council Conservation section, which is usually based in the Planning Department. The heading 'Thatching' in the yellow pages contains entries of local thatchers, and a further register is held by the Information Officer of the Council for Small Industries (COSIRA). Given these sources, you should have no trouble in finding the craftsman whose skills can assure you of continuing cottage comfort.

Clay tiles

Ever since the Roman occupation, tiles of baked clay have been made in this country. It has always been a localised industry, governed by the geological pattern of clay deposit. As a result of regional growth, products from different clay pits had their own unique characteristics. However, the lack of uniformity in size makes repair work difficult, and this prompted Edward IV to draw up a statute in 1477 prescribing a standard dimension. Not all clay pits followed the royal specification of $10\frac{1}{2} \times 6\frac{1}{4} \times \frac{5}{8}$in ($267 \times 159 \times 16$mm), and tiles from Leicestershire were usually much larger. Renewed attempts in 1725, under a statute of George I, reaffirmed the earlier standard. But even if the tile makers agreed to implement a rationalisation strategy, their simple kilns with primitive temperature controls militated against accurate output. Adjustment could be made by over- or under-baking, but this had an adverse effect on the product. Like slices of toast, over-baked tiles become increasingly smaller but also increasingly fragile. Modern technology permits better accuracy, but the five-hundred-year-old standard has not been followed. Current practice is to produce plain tiles $265 \times 165 \times 12$mm ($10\frac{7}{16} \times 6\frac{1}{2} \times \frac{1}{2}$in), with shortened 190mm

Photo 49 Thatching with Norfolk reed commences when gathered bundles are tied down to the rafters and battens

Photo 50 A close-up study of traditional thatching techniques; a task to entrust to the rural craftsman

($7\frac{1}{2}$in) units for the ridge course and the under eaves course.

Until the arrival of modern technology, durability of the end product was sometimes suspect. Errors in the baking process cause early failure, and spalling may be attributable to poor firing. Hand-made clay tiles are also vulnerable for other reasons. With traditional methods, the clay is hand-pressed into its mould, and even if a form of screw press is used, the texture and folds remain visible. This form of roof covering is undoubtedly one of the most attractive, and it is pleasing that hand-made tiles are still obtainable, albeit at a high price. But their performance is not good compared to their machine-made counterpart. For example, the folded layering absorbs and holds moisture and thus it is preferable to use them on steeper roof pitches so that rainwater is shed rapidly. Machine processes overcome this weakness, and the addition of sand to produce a textured surface partly makes up for the loss of hand-made individuality. Moreover, the machine-pressed tile can be used for roofs of lower pitch – for example 40 degrees. With good-quality clay, free from stones and deposits of limestone, and with the correct baking, the durability of the machine-made clay tile is impressive. Past experience has shown that the life of clay tiles can exceed a century, and given the advantages of modern production, one might assume that this longevity will be extended considerably.

High cost is another disadvantage of clay tiles, but in appearance they can be magnificent. They may lack dimensional consistency, but herein lies their charm and character. Indeed it is this individuality and rich texture which has led more and more home hunters to seek old properties for renovation in preference to the purchase of a modern house.

Colour and colour permanence must also be recognised. Regional differences in the composition of clay give contrasting colours. The gault clay in Cambridgeshire yields pale yellow and soft brown tiles. In Kent and Sussex, redder hues provide the mark of regional distinction, whereas in parts of Staffordshire a limited number of blue tiles have been produced. Linked with this benefit is the fact that clay tiles do not fade. On the

contrary, their colour often deepens with age, and this is rightfully regarded as a particularly endearing feature.

Clay tiles were originally attached with oak pegs as described earlier, but in the later 1800s 'nibs' were fitted during manufacture. However, this means of attachment is usually supplemented with non-ferrous 38mm (1½in) nails – especially in exposed situations. For instance, every third course should be nailed on a roof situated on an exposed site and/or where the pitch is steeper than 50 degrees. In less severe circumstances, nailing every fifth course is acceptable.

Modern plain tiles are also manufactured with curvature or 'camber' over both length and width. Essentially this is functional, but it also creates pleasant contrasts of light and shade. The purpose of camber is to reduce capillary action in which rainwater tends to be drawn upwards and sideways.

In the pattern of British building, the plain clay tile has a noteworthy place, even in its simplest rectangular form. It is satisfying without being ostentatious. Some Victorian builders disliked this simplicity, and, in keeping with their liking of ornamentation, produced a fish-scale version. The accompanying photograph (see Photo 52) of an imposing Victorian property shows the result, complemented by chevron ridge tiles and decorative barge boards. But in addition to the 'plain tile', clay is also suitable for making pantiles or variations of the Roman tile. An example is shown of a property in a Northamptonshire village completed in 1690, and it is pleasing to note that clay pantiles are being used in its refurbishment nearly three hundred years later (see Photo 53).

Today, clay tiles represent a very small proportion of the total market – less than twenty per cent. This is partly due to cost, but there is an increasing interest in traditional materials, and an eagerness to renovate old

Photo 53 Reroofed in 1985 with new clay pantiles – a fitting choice of 'traditional' covering material in the refurbishment of a property near Northampton, originally completed in 1690

properties. Hence it is not surprising to learn from The Clay Roof Tile Council that in the early part of the 1980s the usage of clay tiles (as a proportion of all roofing materials) showed a fifty per cent increase. When considering appearance in the context of Britain's heritage of building, one can only feel a sense of delight at the renewed interest in these natural artefacts.

Concrete tiles

The concrete tile is now the most popular form of roof covering. It is believed that the ancient Greeks and Romans knew of concrete, but its use for making roof tiles has a much shorter history. Experiments carried out in southern Bavaria in the 1840s marked the beginning of an industry which soon developed in Britain. At a sandpit at Reigate, Surrey,

Photo 51 Traditional plain tiles made from clay darken with age, and produce a roof style which has long been associated with Britain's building heritage

Photo 52 Decorative ridge tiles, and shaped clay tiles create a striking roof on a property built in the late nineteenth century

Photo 54 These interlocking 'Roman-style' profiled tiles made in concrete are guaranteed for 100 years

Redland produced their first tiles in 1919. The location had not been selected by accident, because one of the important constituents of a concrete tile is fine-graded sand. Redland now manufactures concrete tiles in many parts of the country, but Reigate remains the administrative headquarters. Coincidentally, in neighbouring Kent a builder who was hindered by shortage of materials in the post-war period also experimented with concrete. After successful results, he formed the Marley Tile Company in 1924. Both the Redland and Marley enterprises grew rapidly, and before World War II, the concrete-tile industry as a whole claimed around twenty-two per cent of the total covering materials used for pitched

roofs. It is now claimed that concrete tiles account for seventy-five per cent of the market.

Durability is one of the chief advantages of concrete tiles, and Redland now offer a one-hundred-year guarantee with their products – a warranty which is transferred as ownership changes. No other roofing material receives a guarantee approaching this figure, although quarried slate offers similar long life (*see* Photo 54). When made to British Standard (BS 473 and 550 : 1971 : [1980]), a concrete tile withstands all extremes of temperature encountered in this country. Similarly, the absence of a laminar structure prevents frost damage. Further verification of its strength is its resistance to impact force; the product will not be damaged by large hailstones and is claimed to have resisted 45mm (1¾in) freak examples.

The versatility of concrete is evident in the

array of products. For example, the traditional appearance of plain tiles can be reproduced. Alternatively, tiles can be made with interlocking channels at the sides and a variety of profiles are also available including pantiles and Roman tiles. Surface textures can also be varied. A smooth tile permits more rapid surface run-off and can be laid to a lower pitch. However, a roughened surface may be preferred where colourised granules of crushed stone chippings have been fused to the upper surface. Permutations of colour, shape and texture provide a wealth of alternatives, and to the untrained eye some examples are difficult to distinguish from the equivalent clay product.

In manufacture, the constituents for a concrete tile are Portland cement, clean siliceous sand, crushed ballast and crushed hard stone. This is compressed in pallets, with considerable pressures applied by machinery. Uniform shapes facilitate the roofing exercise, though this geometrical exactitude denies the product the higgledy-piggledy charm of traditional coverings. However, some examples intentionally bear irregularities, such as imitation stone tiles (*see* below). In today's manufacturing process, the work is highly mechanised with a minimum call on labour. Moreover, the concrete tile can be made throughout the year, whereas drying processes in clay-tile manufacture prove difficult in certain seasons. Both factors contribute to the low cost of concrete tiles, which in turn has led to their popularity.

Like the plain clay tile, the concrete counterpart should be laid with a head lap no less than 65mm (2½in) in situations of moderate exposure. With greater weather risk, 75–90mm (3–3½in) is usually specified. Interlocking profiled tiles require a minimum head lap of 75mm (3in), although on low pitches some products require a 100mm (4in) head lap. As head lap increases, the gradient on the face of the tile decreases, and for this reason the lap should never exceed a third of the total length of the unit. Specification details for concrete tiles are clearly stated in the manufacturers' literature, and this information is simple for the 'non-expert' to assimilate. Technical enquiry departments at the head offices of both Marley and Redland provide further back-up.

Although the benefits of concrete tiles are numerous, three disadvantages should be recognised. Firstly, the product is heavier than most covering materials and, whereas this can be an advantage in some situations, the supporting structure must be designed accordingly. However, when wind and snow loadings are taken into account, the increased weight of concrete coverings as a proportion of the total weight is relatively small.

Colour permanence is another shortcoming. Whereas the clay tile darkens with age, the pigmented concrete tile bleaches. When a tile needs replacing, or if a home extension bring new and old coverings together, the mismatch is clearly evident. Mixing new and old units to form a mottled effect reduces the disharmony; manufacturers are also claiming that colour permanence is now much improved.

Lastly, concrete tiles are not easy to cut. Whereas a synthetic slate can be shaped with a cutting knife and snapped over a sharp edge, there is no such ease with a concrete tile. Cutting requires a power stone saw with a Carborundum disc cutting wheel. These machines are easy to hire, and not difficult to use. However, they can be dangerous, and an HMSO guide has recently been published to draw attention to correct usage.

In spite of these criticisms, the list of advantages indicates why concrete tiles are now the most popular covering.

Reconstructed stone

Notwithstanding the advantages of conventional concrete tiles, there are some localities where they fail to blend comfortably with their surroundings. For example in the Cotswolds, where soft coloured stone has endowed great character to the rural landscape, natural stone is the material of roofing. However, it may not be possible to obtain quarried materials, and reroofing schemes can pose special problems if the objective is to preserve existing styles. 'Reconstructed stone' slates often produce the answer.

Reconstructed stone is essentially an imitation material based on a mix of Portland cement and oolitic limestone. If you see the term 'reconstituted stone' this indicates that fragments of the original stone are also blended in the mix. An authentic appearance is also achieved by the moulding process. Moulds are sometimes copies of real stone and their individuality is thereby reproduced in the final product. By manufacturing varying lengths and widths, it is possible to reproduce a random tiling effect and also diminishing courses. One manufacturer, Bradstone, supports the product with detailed literature showing how to achieve this effect. As long as a roof contains simple pitches and large areas rather than intricate details, the task is not too difficult to carry out using the manufacturer's instructions. Variations can also be introduced if you follow the instructions for installing Westwold character roofing, an imitation stone made by the Marley Roof Tile Company.

All the advantages of concrete coverings are again in evidence, but this special moulded alternative, finished in suitable pastel shades, harmonises well with 'real stone'. When mosses and lichens add their contributory colour, the resulting effect will convince all but the expert.

Reconstructed Cotswold stone is particularly suitable for a number of southern counties, such as Oxfordshire, Northamptonshire, Dorset and Wiltshire. However, a grey moorland alternative is eminently suitable for northern counties like Yorkshire, Lancashire, Derbyshire and Durham. Thoughtful development on the part of the manufacturers, and matches in both the colour and style of the original dressed stones, have allowed traditional appearances to be kept alive. To retain individuality in the product, manual methods are still a feature in the manufacturing process, which in turn is reflected in price. Understandably, this is one of the more expensive forms of covering material, but no one would doubt the need to use it on certain buildings.

Slate

It is claimed in the industry that slate can be used for everything in a house except the glazing. Damp-proof courses, steps, window-sills, lintels and coping stones are obvious examples; the four-poster bed made for Queen Victoria and kept in Penrhyn

Photo 55 In North Wales slabs of slate are often used for providing solid anchorage at verges. But this technique leads to the ingress of rainwater from capillary action

Castle near Bangor is a less obvious example. This valuable natural resource is available in several parts of Britain. In Wales it has been quarried commercially in at least half a dozen counties. In England slate has been quarried in Devon, Cornwall, Leicestershire, the Lake District and the Isle of Man. Long recognised as a roofing material, slate from the Swithland quarries in Leicestershire was used by the Romans, and in the twelfth century slate from Devon and Cornwall was shipped around the country.

Slate varies considerably according to its place of origin, and it is somewhat unfortunate that prolific inner-city building work during the last century has conveyed the idea that slate is dull. On the contrary, Welsh slates display notable variations in colour and texture, even from different levels within the same quarry. Welsh slate can also be shaped easily, which adds opportunity for a versatility rarely exploited in volume building. Without doubt the slate in the villages of northwest Leicestershire or the rough and smooth textures in the Lake District are the antipathy of monotony. Often the beauty of the material is evident nearer its source, and pattern effects using slates of contrasting colours are cleverly executed in parts of North Wales. The French exploited shaping, and the striking fish-scale effects on the turreted towers of their *châteaux* is *de rigueur* abroad but less common in England. It is a practice, however, which can often be seen in Scotland and in parts of Wales. Another strategy is to use slates of random widths, laid in diminishing courses as shown in Photo 56. To install a roof in this manner is a skilful task, and an error in bonding leads to leaking. However, it again shows that slate roofs are not as 'pedestrian' as one might suppose.

The slates of the Lake District have more attractive surface textures than most Welsh slates. However, this prevents them from bedding down closely and increases the problem of wind lift. On the other hand, their extra weight helps to combat wind problems, though it imposes greater demands on the supporting roof structure. Welsh slate can be split into extremely fine and smooth-faced pieces which gives an important weight advantage. The consequent ease of bedding down in close proximity reduces the likelihood of damage from wind lift, and Welsh slate can be used for roofs of much lower pitch, for example 22–24 degrees. In contrast, roofs built of English slates usually require a pitch no shallower than 30 degrees.

Strength, durability, minimal porosity and resistance to wind-borne pollutants are features which make slate an ideal roofing material. For example, spalling is not a problem with Welsh slate. As the industry grew, the establishment of standard slate sizes became the practice, and these were given unusual names. Typical examples would measure (quoting their original imperial units and present day metric equivalents) 16 × 8in, 20 × 10in, 24 ×12in and 30 × 24in (405 × 205mm, 510 × 255mm, 610 × 305mm, 760 × 600mm), and these bore the whimsical titles of 'ladies', 'countesses', 'duchesses' and 'Bangor queens'. This quaint practice will undoubtedly disappear with metrication. The metric dimensions of slates are set out in BS 680 : Part 2 : 1971 and BS 690 : Part 4 : 1974 where no fewer than twenty-seven standard sizes are listed; with reason one might mourn the loss of the ladies.

Slates are also graded according to thickness, strength and their degree of porosity. However, the traditional terminology used to describe thickness is likely to be confusing. For example, if you hear reference to 'bests' or 'firsts', this is not a qualitative measure but simply a term used for the thinnest slates taken from a quarry. Similarly, thicker slates are called 'seconds' or 'thirds', but in no respect does this imply any reduction in standard. If you are replacing broken slates on an existing roof, thickness as well as length and width dimensions must be taken into account. A demolition contractor may be able to supply slates for repairs, but you must check the thickness of the replacement units.

In the country as a whole, the widespread

Photo 56 Slate laid in random widths and diminishing courses is a skilful form of roofing technique best left to the professional

use of slate for roof covering has steadily declined. Alternative coverings have grown in popularity, and artificial slates are often used in preference. It is certain, however, that these are unlikely to match the longevity of natural slates, whose life may exceed two hundred years, and on average reaches at least a century. Neither will artificial colourings nor the resistance to frost match the performance of the genuine article. But cost is often the criterion of choice, and covering a roof with new Welsh slates may work out at twice the cost of the synthetic-fibre competitor.

As a final note, it is important to mention that Spanish slates have recently been imported from Galicia. These are available at competitive prices, but in quality some examples do not comply with British Standard 680, or satisfy local authority requirements. One

reason for the interest in a foreign product has been delivery delays of Welsh slate; however, if you consult the Penrhyn Quarry telephone help-line service in North Wales, you will find that many sizes are available ex stock, and only large units are likely to incur delays in supply. Recognising that slate is not difficult to install on a simple roof, this will be of particular interest to the amateur builder.

Synthetic-fibre-bonded cement slate

Natural slate has long been associated with roofing. However, synthetic slates made from cement and bonded with a fibre reinforcement have become increasingly popular. Until recently, asbestos provided the reinforcing bond, and it was a temporary blow to the industry when this was deemed a health hazard. Synthetic fibres and filling compounds have now been introduced as an alternative to asbestos, and manufacturers' brochures often bear the designation 'non-asbestos formulation'.

Photo 57 Synthetic slate, made from compressed fibre, cement and colouring pigment, gives a distinctive roof covering which is easy for the amateur to install

One of the advantages of cement-fibre slates is their light weight. This may be an important factor in refurbishment work where a timber-roof structure is best served with a light covering material. Variety in shape is another factor to recognise; one manufacturer offers five 'wavy shaped' options in addition to rectangular units. Fish-scale and other reticular patterns can thus be contrived without difficulty. Since the product is moulded, several manufacturers include matching ridge tiles, ventilating units and gas vents in the range. In respect of fixing, the artificial slate can be supplied with holes already formed, which is a useful time saver. Clipping systems are also available, but are better suited to the contractor. Shaping, however, is notably easy, and slates can be scored with a cutting knife and then snapped over a sharp edge. Lastly, price is appealing, although comparisons usually reveal that the synthetic slate, though cheaper than quarried slate, is more expensive than concrete tiles.

Notwithstanding the benefits of fibre-cement slates, life expectancy falls far short of the quarried product. Manufacturers give a guarantee of around thirty years against disintegration, and only time will tell if this will be significantly bettered. Early examples suffered badly in respect of colour loss, and washed-out roofs are not uncommon. Recent products are claimed to be better, and it is likely that pigmentation permanence will gradually improve. Two methods are generally used for adding colour; some slates are surface coated with adherent colourings, whereas others are pigmented through the whole tile. When the slates are new, there is no doubt that the result is distinctive as Photo 57 illustrates.

Plastic concrete slates

Modern chemical technology has now led to the introduction of 'plastic' slates which bear close resemblance to the quarried product. One example bearing the trade name Anglia slate is described by the manufacturer as 'methacrylate polymer concrete'. With anticipated minimum life expectancy of fifty to eighty years, this 'look alike' synthetic slate can even be mixed with natural slate without discomfort. In the moulding process, rough-cut edges are created, and the irregular surface texture looks remarkably like the 'genuine article'. Care is taken to reproduce natural colours, and for quantity orders the manufacturer even operates a 'colour to order' service if matching is an important consideration. The product has many advantages such as its resistance to air-borne pollution, and durability in adverse weather conditions. In the long term, an unusually high flexibility and tensile strength will undoubtedly reduce breakages to a minimum. Moreover, its resinous surface provides an inhospitable home for lichens and mosses. Four sizes of Anglia slate are available, and pre-drilled holes facilitate speedy fixing.

Another interesting product is a reconstituted slate from Redland which was launched nationally in 1986. The product, known as Cambrian Slate, is made with sixty-five per cent slate granules reinforced with fibreglass in a resin binder. The units are moulded to reproduce the familiar riven edge of natural slate, but are mounted within a frame which features interlock channels. When installed, the covering bears close resemblance to quarried slate, as well as offering the advantages of a single-lap product, which are discussed on page 113.

In the last decade, plastics have been used increasingly in building work, and these products look most interesting.

Galvanised-steel profiled tiles

This unconventional covering, imported from abroad, has the particular advantage of being able to withstand unusual temperature extremes; one of the leading types is known as Decra tiles. More important in this country, however, is the fact that these coverings can be laid on pitches of 15 degrees, and even 12 degrees if special precautionary measures are taken. As regards appearance, the product is not dissimilar to profiled tiles made from other materials, and a pseudo-pantile effect is created in the moulding of the steel core. On the galvanised-steel base, a coating of bitumen emulsion is added, followed by granules of natural stone. Often the units are supplied in three-tile widths, and a sprayed acrylic glaze provides the top surface with good protection. Installation follows tradi-

tional methods, and literature detailing fixing procedures is clearly presented. Compared to concrete tiles, a guaranteed life of twenty years is short, although it is the surface coating which is likely to deteriorate first.

Other coverings

It is not possible to present a detailed review of every type of covering because there are so many. For instance western red cedar shingles are omitted because they are seldom used on domestic premises. Sheet-metal

Photo 58 Completed recently, this roof covering of modern profiled metal sheeting represented a departure from our more traditional materials

roofs are also omitted and the bitumen tiles mentioned in Chapter 2 are considered more suitable for recreational buildings. And lastly there is no description of the Platon system for turf roofs which is now available in this country from Planox Ltd. However, in case you decide to build a Norwegian-style cabin, with a lawn on the roof, the importer's address is included in Appendix 2 on page 239.

6 Pitched Roofs – Installing a New Covering

The feasibility of installing a new covering yourself depends on access, the style of roof and the choice of material. There must be no compromise on safety, and as long as the roof design is simple, many coverings lend themselves to installation by amateurs. The smart porch shown in Photo 59 represents an ideal first-time project for the self-builder.

The importance of seeking expert advice on all building projects apart from simple sheds has already been stressed. An architect's drawing will specify the covering material, and although you may have discussed several options with him, the final decision will have required local authority approvals. In refurbishment schemes, re-

Photo 59 Simple roofs are likely to pose no problems for the amateur; this smart porch represents an ideal introductory project

claimed materials may be preferred. If local roofing contractors cannot help with unusual items, the Consumers' Association reported on dealers who supply second-hand building materials. ('The Good Salvage Yard Guide', *Which?* magazine, May 1985.)

For information on new materials, it is often fruitful to approach the manufacturers direct. Some operate Freephone advice services and most are exceptionally good at providing product literature with supporting notes on installation procedures. Using the explanations in this manual, you should have no difficulty in following their technical leaflets. Some companies will then refer you to a regional sales manager or a builders' merchant according to their supply system. With large companies, a rep may call on site, take measurements, prepare a listing and submit a quotation. You are advised to use this service if it is available; calculating quantities and detailing every small item are not easy. Check delivery dates carefully; there is no point in hiring scaffolding until the tiles or slates are available.

The underlay

In both new work and refurbishment schemes, current practice considers it essential to provide a secondary barrier or underlay. This serves a number of important functions described in British Standard Code of Practice (BS 5534 : Part 1 : 1978). The underlay represents an additional barrier to wind, rain and snow; it also acts as a thermal and sound insulant, and helps to keep dust out of the loft space. The cheapest barrier is formed with a purpose-made roofing felt, but in Scotland and exposed areas noted for bad weather, roofs are boarded as well as felted. The only disadvantage of an underlay is the fact that it prevents inspection of tiles or slates from the loft.

Bitumen sarking felt

For a number of years this has been the most popular material, but you must buy the correct type of felt. Roofing felts are classified in British Standard 747 (1977), and the one we are concerned with here is reinforced on one side with an embedded layer of jute hessian. With this reinforcement, it is far less likely to tear compared with bitumen felts used for flat roofing. Always check, however, that the reinforcement is on the underside. It is worth adding that for properties where heat loss is a problem, you can buy felt coated with aluminium foil, although this precautionary measure is not often taken.

For the most part, roofing felt is held in place by the roof battens which are laid on top of it. However, initially it should be fixed with galvanised iron clout-head felt nails,

taking care to select a day when there are no strong winds (*see* Fig 55). The felt is laid across the rafters parallel to the eaves, starting at the lowest point. Spare overhang – about 75mm (3in) – should be left projecting beyond the fascia boards, and this is needed to hang over the runs of guttering. To prevent the lowest tile from hanging at too low an angle, the upper edge of the eaves fascia board will have been fixed above the rafters. This can cause an upturn in the felt, and a chance for rainwater to accumulate. The problem is less likely to occur if sprocket pieces are nailed on the lower portion of each rafter to elevate the felt (*see* Fig 47, page 85). An even better arrangement is to fix exterior-grade ply 300mm (1ft) wide across the whole width of the roof. Similar support is provided with Redland RedVent eaves ventilators (*see* Photos 96a and b, page 162).

Working up the roof, overlap each horizontal join by 150mm (6in), although for pitches steeper than 35 degrees a 100mm

Fig 55 Requirements when installing a sarking felt underlay

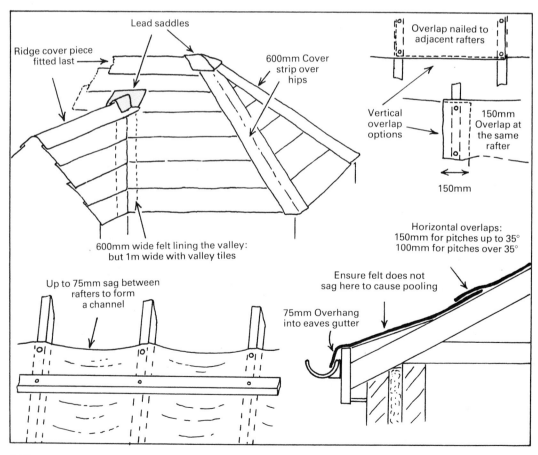

(4in) lap is acceptable. When you get to the end of a roll, it is best to make vertical laps sufficiently generous to allow the separate pieces to be nailed to adjacent rafters. Alternatively a 150mm (6in) vertical lap can be formed at a single rafter. At the ridge, a length of felt is used to drape over both sides of the roof slope. On a hip roof, lengths of sarking felt, cut 600mm (2ft) wide, are laid over the ridge of the hip. If the roof is designed with valleys, lengths of underfelt should be laid first as shown in Photo 60. Strips at least 600mm (2ft) wide are needed, and these will be covered by the felt used for the roof faces on either side. When the entire roof is covered, the property will have a small degree of 'rainproofing', but battening should follow without delay on account of vulnerability to wind. Obviously there have to be some breaks in the continuity of felt for soil pipes or ridge vents, and careful cutting with some old scissors ensures a good fit.

When laying felt, it is essential to allow it to sag slightly between each rafter (*see* Fig 55). If rainwater does force an entry, it is then channelled underneath the battens, and discharged into the eaves gutter.

Photo 60 When preparing valleys, the installation of roofing felt, overlapped generously with horizontal felting, is an essential provision

Boarded roofs

Cost is the main reason why boarded roofs are not commonly specified for domestic buildings, except in exposed areas. In old buildings, unplaned softwood was used; typical material was 19mm (¾in) thick, and between 150–225mm wide (6–9in). Today, tongued and grooved planed board is preferred, and this gives a solid and continuous barrier.

On older properties roofed in slate, the units were often nailed direct to the boarding; this practice is no longer recommended. Moreover, in modern roofing, the boarding must also be covered with a special breather felt which prevents condensation forming on the timber. Prior to fixing horizontal battens, a boarded roof also needs counter battens as shown in Photo 61. This ensures that if rainwater gets underneath the slates or tiles, it will have an unimpeded flow to the eaves gutters. There is no need to use material of any notable size or strength for this purpose, but

111

the cross-section of counter battens should not be less than 38 × 6mm (1½ × ¼in); in practice, sturdier material (eg 38 × 25mm, 1½ × 1in) is often preferred, especially if this is already on the timber order for the main battens.

Alternative underlays

Recent developments include insulating materials incorporated into underlays, and the introduction of plastic sheeting as an alternative to bitumen felts. The latter are lighter to use and very resistant to tearing. The main shortcoming of plastic underlay is the tendency for condensation to form on its underside, although tiny perforations help to improve permeability. It would be no surprise to see this alternative to bitumen felt becoming more popular in the future.

Photo 61 In Scotland roofs are often boarded, felted, and covered with 'counter battens' before horizontal battens are installed. This secondary barrier offering a high level of weatherproofing is shown on a housing development near Aviemore, Inverness-shire

Tile and slate battens

Both tiles and slates are usually fixed or hung on horizontal battens, which in turn are fastened to the rafters. Since many covering materials are expected to last in excess of fifty years, it is logical to ensure that the battens on which a covering is laid will enjoy similar longevity. For this reason, battens should be purchased which are impregnated against rot, as laid down in BS 4072 : (1974). However, a live issue at present is that the copper-chrome arsenic used as a batten preservative is believed to corrode aluminium roofing nails. The Building Research Establishment is engaged in a study of this problem. At the other extreme, it is not unusual to see roofing work which uses untreated softwood for battens.

The dimensions of battens are specified by the manufacturers of roofing materials according to the spacing of the rafters and the weight of the tiles or slates. Recommendations are based on data tabled in the May 1982 amendment (AMD 3554) to BS 5534 : Part 1 : (1978). This is shown in

Appendix 1 (Table 2) and indicates the different provision for 450mm (18in) and 600mm (2ft) rafter spacings. It should also be appreciated that in the design of trussed-rafter roofs, battens represent an integral part of the structure, and act as bracings to prevent deflection in the trusses. Since battens provide attachment for the covering units, anchorage for the sarking felt and support to the structure, the need to use material of the recommended dimensions is particularly important.

Fixing tiles and slates – two methods of installation

Most pitched roofs are covered in either tiles or slates, but these can be divided into two categories: namely, single lap or double lap. The difference determines the way in which they are installed, and thus it is appropriate to look at the two methods separately.

Single-lap roofing

Single-lap fixing is employed for coverings which incorporate a moulded interlock groove at the sides. Profiled units like modern pantiles, Roman tiles and trough tiles are examples, and it is immaterial whether these are made of concrete or clay. Concrete flat tiles with grooved side-locks have also been introduced, (eg Redland's 'Stonewold Slate' and the Marley 'Modern'). In single-lap work, the head of a tile is covered by the tail of the tile in the course above it, but there is never more than one tile overlapping another. This is shown in Fig 56. Interlock at the sides also provides weatherproofing, and only in the case of flat tiles is the groove staggered in adjacent courses. Aesthetically this is preferred, and staggering gives a bonding effect familiar on roofs covered in quarried slate. In terms of performance the staggering prevents rainwater tracking down the interlock. In contrast with most profiled tiles, the groove represents the lowest point on a flat tile, and hence the problem.

The principle of interlocking has a number of advantages and one of the most recent developments has been the introduction of a reconstituted slate mounted in an interlock framework. 'Cambrian Slate' from Redland

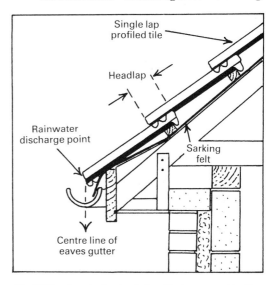

Fig 56 The head of a single lap tile is covered by the tail of the tile in the course above it. This gives double thickness at each head lap as shown with these modern Roman tiles

is a close replica of quarried slate, but it is installed as a single-lap roofing material on account of the interlock innovation.

Single-lap tiles are straightforward to install, and well suited to the beginner builder. Dry-fix versions are even easier, and a growing wealth of literature from manufacturers gives clear instruction on procedures. However, some roofs do not lend themselves to interlock tiles as we discussed earlier. For example, they are not suitable on steep pitches or on roofs with intricate detailing. In the descriptions which follow, it is presumed that your roof is of simple design and suitable for single-lap units.

Establishing the position of the battens

Forethought in the positioning of roof tiles will eliminate unnecessary cutting at abutments, soil stacks and roof windows. You must always commence work by laying out the eaves course of tiles, although this is done as a dummy run before making a final fixing. Eaves tiles must be positioned so that the lowest edge from which rainwater discharges is directly above the centre line of the eaves guttering. Since guttering is not added until later, verify its position by temporarily offering up a bracket to the fascia board (*see* Fig 57). In the case of a standard 100mm (4in)

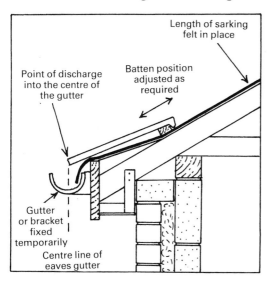

Fig 57 To work out where to nail the bottom batten, temporarily fix a bracket or short length of gutter to the fascia board. Then offer up a batten off-cut and an eaves tile, adjusting their position until the point where rainwater would discharge is directly above the centre line of the gutter

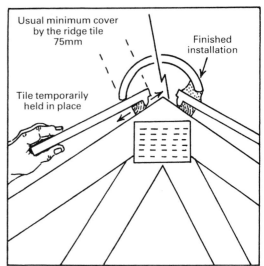

Fig 58 With the top batten nailed temporarily, its position can be verified by offering up a top course tile. If the ridge tile is held in place next, it should provide a minimum head cover of 75mm (3in). Altering coverage is done by re-positioning the batten

eaves gutter, the lower edge of the tile usually extends 50–65mm (2–2½in). The position of the eaves tiles dictates the position of the lowest batten, and when sarking felt has been fixed this can now be nailed into position. However, no nails should be driven within about a metre (3ft) of the verge because an undercloak will later have to be slid under the end of the battens. The undercloak provides a platform for mortar which is used to seal the verge.

The position of the top batten is dictated by the covering capacity of the ridge tile. If a short length of trial batten is temporarily nailed in position, a tile placed on it and the ridge tile offered up, the specified minimum head lap should be covered by the extremities of the ridge tile (*see* Fig 58).

Given the position of the top and bottom batten, and assuming the roof has been built accurately with equidistant spacing from ridge to eaves, you should now calculate the number and spacing of the intermediate battens. It must be remembered that the gauge of the battens is the same as the margin, ie the amount of tile showing. Moreover, the gauge is the measurement from the centre line of one batten to the centre line of the battens above and below. Maximum permissible

gauge is generally listed in manufacturers' specifications, and is worked out as:

$$\text{Overall length} - \text{minimum} = \text{maximum}$$
$$\text{of tile} \qquad \text{lap} \qquad \text{gauge}$$

Gauge and lap are wholly inter-related; if you decrease the gauge and bring battens closer together, the head lap of the tile will increase proportionally. Putting this into practice, the distance between the top and bottom battens has to be measured first. This should then be divided by the maximum gauge permissible in order to establish the number of spaces between battens. It is highly unlikely that the arithmetic will work out exactly, and the answer must be increased to the next whole number. Since this represents the division of the roof into equal measures, a subtraction of one will indicate the number of battens required between the eave and ridge battens already located. An imaginary step-by-step example illustrates how this calculation is carried out.

Example
1 A typical set of data provided by a manufacturer of profiled interlock tiles might read:

Overall size of each unit – length 413mm (16¼in); width 330mm (13in)

114

Minimum lap – 75mm (3in)
Maximum gauge – 338mm (13¼in)
Lap can be increased up to a maximum of
100mm (4in).

2 It is found that the distance between the
centres of the bottom and top batten measures 3.500m (11ft 6in).

3 Maximum gauge quoted by the manufacturer – 338mm (13¼in).

4 3500 (138) divided by 338 (13¼) = 10.35
(10.41) spaces between battens.

5 Increasing this to the next whole number
gives eleven spaces, ie ten more battens are
required between the two already temporarily *in situ*.

6 Having made this adjustment, the gauge
will be decreased accordingly, albeit within
the permissible limits, and the lap at the head
of each tile increased.

Given the number of battens required, you
are left finally to calculate the actual gauge,
so that battening can be nailed into position.
Dividing the measured distance between top
and bottom battens (3.5m/11ft 6in) by the
number of spaces (11) = 318.18mm
(12½in). For practical purposes, the roofer
does not work in fractions of millimetres, and
the battens would therefore be spaced at
318mm (12½in) gauge. Occasional checks
must be made every three or four courses to
establish if any accumulated error deserves
slight adjustment.

These calculations show how to divide up a
roof to accept battens at equal spacing, and
where head lap adjustments have been a
mathematical 'rounding-up' exercise, dictated by roof dimensions. The procedure assumes that the building has no undue level of
exposure and will not be subjected to unusual
weather extremes. In doubtful situations,
particularly if the pitch approaches the lower
recommended limit, head lap may need to be
increased. This reduces the margin, adds
weight and counters wind lift. However, increasing head lap has the effect of decreasing
the pitch on the face of each tile. In consequence the dispersal of rainwater is reduced,
and there is a greater likelihood of moisture
creep. As a rough guide, head lap should

never exceed one-quarter of the overall
length of a profiled tile, and manufacturers'
recommendations are usually less than this.

If bad workmanship in the roof structure is
revealed in distance variation between the
ridge and eaves battens, errors should be rectified by adjusting the cover of the ridge tile.
Adjustment must not be made at the eaves
on account of the need for accuracy in the discharge of rainwater into the guttering.

Battens are usually fixed at the same time
as the felt is fitted. For quickness you can cut
a couple of spacers from an offcut as shown in
Fig 59. Spacer length will be gauge minus batten width. Measuring out batten positions in
this way eliminates the need for constant reference to a tape measure. However, a
periodic tape check is recommended to make
sure that small errors do not accumulate.

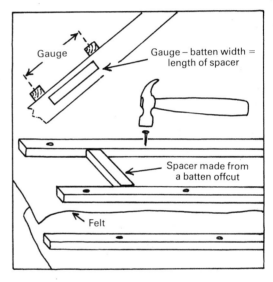

Fig 59 Battens can be positioned with the help of a
spacer made from an offcut

As felt is unrolled and battens positioned,
a useful 'batten ladder' is thus formed at the
intersection with the rafters. It is safe to place
weight at these points, but *not* between the
rafters.

Battens are unlikely to be long enough to
span a roof; joints made at the rafters should
be formed by skew nailing, and not more
than one batten in every four should be
joined on a rafter where trusses are used. Fixing requires 10 gauge galvanised wire nails at
least 32mm (1⅜in) longer than the thickness
of the batten. Any course-toothed saw is suit-

115

able for trimming battens at verges and abutments as shown in Photo 62. On a hipped roof, you will need to install battens on either side of the hip rafter as shown in Photo 63 and 69a.

Positioning the eaves course and marking out

As a preliminary, the eaves course is laid out temporarily; on a duo-pitch roof, the right-hand verge tile should overhang the gable end by about 40mm (1½in). On reaching the left-hand verge it will become apparent if the last tile overhangs to a greater or lesser amount than 40mm (1½in). After adjustment, verge overhang should finish between 38 and 50mm (1½ and 2in) at either end. The whole course can be realigned, but you also have another means for manoeuvre, known as 'side-lock shunt'. This is the chance to open up or close down each interlock groove. It may only amount to 3mm (⅛in) per tile, but over the complete course this tolerance accumulates to provide useful leeway. On a hipped roof, cutting will be necessary so that the end tiles of each course align with the rake of the hip.

Once the layout of the eaves course is established, you should then mark the battens to ensure that the interlock joins sit in perpendicular alignment. A traditional method is to use string lines coated in cement dye or ochre which are stretched up the roof tautly, and 'pinged' so that a deposit mark is left. Pencil lines are acceptable but take longer to mark. A perpendicular can be formed with a home-made ply square as shown in Fig 60. If its sides are measured out in 3 × 4 × 5 proportions, a right angle is formed opposite the longest side. An extended string line will carry this perpendicular from eaves to ridge. However, before the eaves tiles can be fixed permanently, an undercloak is required at the verges.

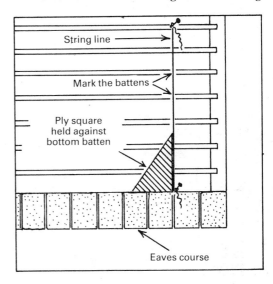

Fig 60 Perpendiculars should be marked out on the battens using a string line coated in either ochre or cement dye

Fitting the undercloak

An undercloak can be made from a variety of materials, but when interlock profiled tiles are used, cement/fibre strip is a common choice. This must be at least 4.5mm (³⁄₁₆in) thick, and pieces 150mm (6in) wide fulfil the requirement. Cement asbestos strips were formerly used, but for health reasons a substitute is now preferred. The shiny side should be laid downwards, and prepared lengths are eased underneath the battens at the edge of the gable. Sarking felt should be trimmed to allow a 25mm (1in) excess to be carried beneath the undercloak. When the board is laid directly on to brickwork, a thin mortar bed is required for support. If nailed directly to a timber base, for instance a flying gable, the mortar is unnecessary. You will find that cement-fibre boarding is fragile, but pre-drilling will prevent the galvanised nails from splitting the material. Check that the projection of the undercloak equals the projection already established for the verge tiles. Once positioned, nails left out of the tile battens to leave space for the undercloak should now be driven home.

An undercloak can also be formed from slates or plain tiles laid face downwards and bedded on mortar. A temporary batten nailed to the gable provides support while the mortar sets, and helps to align each tile. This

Photo 62 Battens are trimmed to length at verges, hips and abutments. Any coarse toothed saw is suitable, although special saws are available for roofing work

Photo 63 Once calculations for gauge and lap have been worked out, felt and batten work is a straightforward operation. On a hipped roof, battens are installed on both sides of the hip rafter

117

Photo 64 The verge – exposed to show the cement/fibre board undercloak in position below the battens. Note the sarking felt has been cut rather short, thus losing its lap with the undercloak

is an attractive alternative, but it takes much longer to complete.

Fixing tiles

The nibs on moulded tiles are an effective means of attachment, but mechanical fixing may also be needed. This is to resist wind lift, and you must use nails or clips in accordance with the manufacturers' recommendations. It is unusual, however, to fix every tile, unless the pitch is steeper than 45 degrees or the site unduly exposed. A more usual procedure is to nail or clip the eaves course, the ridge course and every third course in between. Nails or clips must be non-ferrous metals, and aluminium, copper or silicone bronze are recommended. Aluminium is usually preferred, and is rather easier to obtain. There

should also be mechanical fixing for all tiles at the perimeters, such as the verges, the sides of valleys and hips, and tiles meeting with an abutment. With regard to the verges, you may find that some manufacturers only require the left-hand verge to be fixed. This acknowledges that the right-hand verge tile is held down at the interlock by its neighbouring unit.

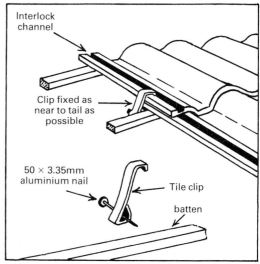

Fig 61 Clips are sometimes used as an alternative to nail fixings

As an alternative to nail fixings, special clips are manufactured which are hooked on to the interlock groove and nailed to the sides of the batten (*see* Fig 61). At left-hand verges, a different metal clip is used which is visible on the face of the tile as shown in Photo 65. These are necessary because the mortar used at verges only acts as a filler, and often cracks through shrinkage. Its adhesion between the tile and undercloak provides little if any anchorage at a point particularly exposed to the wind.

Photo 65 Finished work showing the left-hand verge laid on brickwork, with undercloak, bedding mortar, and fixing clips

Photo 66 To keep out birds, bees and vermin when profiled tiles are used, Marley has introduced a comb pattern filler unit. This provides closure at the eaves for tiles in a variety of patterns. Note also the longitudinal vent which is part of the Marley ventilation system

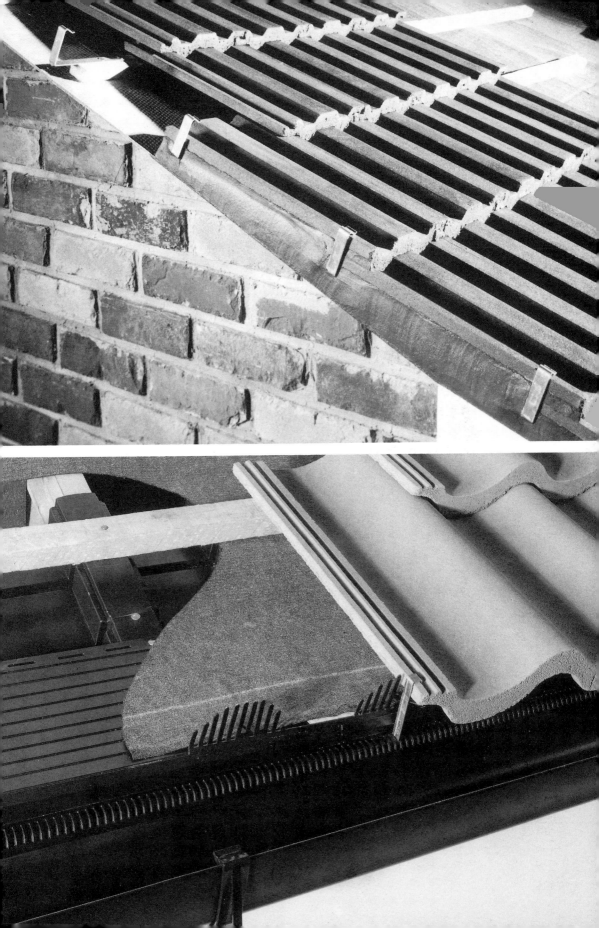

A ◄

C ◄

◄ B

Photo 67a Colourised mortar is distributed in generous quantities along the outermost edge of the undercloak

Photo 67b On the left-hand verge, where the last of these interlocking tiles has no neighbouring cover unit, aluminium clips fastened to the battens effectively combat wind lift

Photo 67c A small amount of mortar is added at the top of the interlock groove to provide bedding at the point of the head lap

Photo 67d Surplus mortar is removed with a round-headed trowel, prior to the final pointing up

D ◄

Eaves fillers

Before the eaves course is finally fixed, apertures under the rolled shape of pantiles and Roman tiles must be closed. These represent an ever open door for birds and vermin, and purpose-made filler pieces are obtainable in both aluminium and plastic to suit different tile profiles. Sections have to be nailed to the uppermost edge of the fascia board in order to coincide with the chosen tile positions. A versatile filler unit introduced by Marley Roof Tiles is made from flexible plastic strips which look rather like a comb set on edge. Its installation is clearly shown in the photograph (*see* Photo 66).

Fixing verge tiles

After fixing the eaves tiles, the right-hand verge tiles are tackled next. Verge-fixing procedure is shown in the accompanying sequence of photographs (Photos 67a-67d). The mortar is made with three parts of sharp sand to one part of Portland cement (by volume), with clean water added to form a stiff mix. You can add mortar plasticiser to make this more 'creamy', although the use of masonry Portland cement achieves the same objective. Often a colouring pigment is added, and powders can be obtained from builders' merchants. These should comply with BS 1014, and you must exercise care in mixing to maintain colour consistency. It is recommended to set the verge tile to tilt slightly inwards, al-

though this is more important with plain tiles; the shape of profiled tiles already prevents a discharge of rainwater over the sides of a roof. Once positioned, the excess mortar will be struck off and then smoothed to produce a good finish. A blunt-nosed pointing trowel is best for this operation.

The over/under of the interlock grooves means that each course of tiles is laid from right to left. At the left-hand verge the last tile will have its interlock groove exposed; some manufacturers, for example Redland, make a special 'left verge tile' which gives a smarter finish.

Cutting tiles

It is often necessary to cut tiles to fit around a chimney stack, or to align with the rake of a hip. Concrete tiles pose a harder challenge than clay, and you will need to obtain an abrasive Carborundum disc machine from a tool-hire shop. These machines can be very dangerous, and if a disc flies asunder under

Fig 62 Ridge tiles are made in a variety of shapes and styles.
A Third round segmental ridge tile (Marley)
B 'Modern' ridge tile for Marley 'Modern' and Wessex coverings
C Marley segmental mono ridge tile
D Roll top ridge tile (Red Bank)
E Cocks comb crested ridge tile (Red Bank)
F Plain half round hip stop end ridge tile (Red Bank)
G Capped angle ridge tile (Red Bank)
H Plain 'T' piece junction ridge tile (Red Bank)
I Slotted roll top ridge tile to accept specially shaped insets (Red Bank)

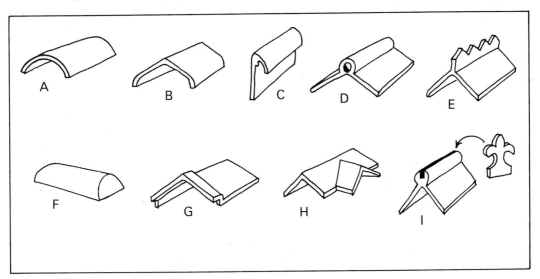

full speed, serious injury can occur. Industrial gloves and eye protection *must* be worn, and the arms should be covered. It is a good idea to insist that the hirer shows you exactly how to hold and operate the machine.

Ridge tiles

Ridge tiles are manufactured in a number of shapes to suit different coverings and various roof pitches (*see* Fig 62). Half round, angular and hog's back are the more common types. On a mono-pitch roof, a special ridge tile is also available. Hips can be covered in a similar way using a segmental tile. Cutting a tile or two may be necessary to match the length of a ridge, and some forethought should be given to decide where a reduced tile or tiles would look neatest. The Carborundum disc cutter already described is the machine to cope with this work.

Installation is shown in Photo 68, and you will need mortar for bedding and jointing consisting of three parts sharp sand and one part Portland cement. To offer support to the mortar, use remnants of broken tile to bridge the gap at the peak of the roof. This is par-

ticularly desirable when a roof has been constructed with roof trusses, since these lack the ridge board or 'ridge tree' used in traditional construction. On reaching the end of a ridge line, the final tile will require a substantial mortar fill. To reduce the likelihood of shrinkage, pieces of plain tile should be inset and pointed up neatly. Not unusually, these sections are left exposed and made into a decorative feature.

Hip tiles are laid in a similar manner, as shown in the photographs (*see* Photos 69a-69e). Commencing at the lowest point, a galvanised hip iron must be nailed to the base of the hip rafter, to offer mechanical support. You must then cut the bottom angles of the tile with a disc machine to align with the lower edge of the adjacent eaves tiles.

When a profiled tile is used, its trough will inevitably take a greater amount of mortar along the side of the ridge units. This is usually reduced by using small 'fingers' of tile which rest in the base of the trough, held in place with mortar. These are known as 'dentils' or 'dentil slips', and nowadays they are purpose made (*see* Fig 63). In older roofing practice, small pieces of plain tile would be used instead, and these were referred to as 'gallets'. Careful pointing is again necessary, and in older roofing a feature was made of the pattern effect created by the exposed ends of gallets.

Photo 68 A rounded trowel is used to point up on a traditional 'wet' mortared ridge. The mortar fill is carefully finished to conform with the trough on these profiled tiles

Photo 69a Felt and batten groundwork shown in close-up, prior to cutting the last tiles in each course to conform with the rake of the hip

Photo 69c Progressing up the ridge, hip tiles are laid successively on a sand/cement mortar bed

Photo 69b A bedding of sand and cement mortar is generously placed around the periphery of the hip tile unit

Photo 69d Pointing up the edge bedding of the hip tile and its butt joint completes the operation

Photo 69e Close-up detailing of a hip constructed with plain tiles and a segmental ridge cover unit. Note the lowest hip tile has been cut to match the eaves corner, and the galvanised hip iron provides mechanical support

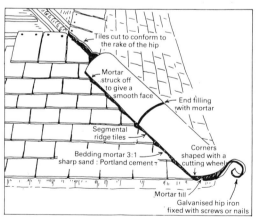

Tiles cut to conform to the rake of the hip

Mortar struck off to give a smooth face

End filling with mortar

Segmental ridge tiles

Corners shaped with a cutting wheel

Bedding mortar 3:1 sharp sand : Portland cement

Mortar fill

Galvanised hip iron fixed with screws or nails

Photo 70 Ridge detailing showing the shorter 'tops tile' at the final course, and a Marley Ridge tile with attachment wire for producing a mechanical fixing

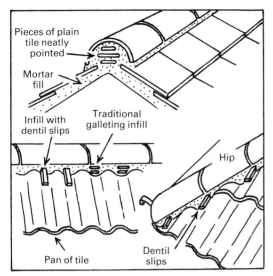

Fig 63 Pieces of plain tile inset into the end of a ridge unit reduce the problem of mortar shrinkage, and produce an attractive finish. In the trough or 'pan' of profiled tiles, galleting achieves a similar function, though nowadays, purpose-made dentil slips are usually preferred

Fig 64 Redland 'Universal' Valley Trough tile. An angle cut on the lowest tile has to be made so that it can be aligned with the fascia boards

Irrespective of workmanship, mortar cannot provide sound anchorage for ridge tiles on exposed roofs. Whereas it is an effective filler, it lacks tensile strength. In particular, the end tiles of a ridge are often dislodged in winds and require frequent refixing. The only satisfactory answer is to use a mechanical fixing – another plus point for the new dry-fix systems. Marley overcome the problem by supplying an attachment wire with their ridge units as shown in Photo 70. On a traditional roof, this would be wound around nails driven into the ridge tree; on a trussed roof, a supplementary top batten is needed to provide anchor points. Either way the result is most favourable.

Ridges may incorporate special flue vents for gas appliances and ridge ventilation units, described in Chapter 7. Fixing procedures are accurately detailed in manufacturers' product information sheets. The work involves cutting an aperture in the sarking felt, and, on a traditional roof, part of the ridge board must be removed.

Valleys

Valleys can be constructed using a variety of materials. However, with a profiled tile, an 'open valley' is required, which is essentially a drainage gutter, 100–125mm (4–5in) wide. This can be formed with troughed valley tiles, a pre-formed channel of rigid plastic, or by lining a box structure with metal. Roof pitch may be a determinant of the type selected, and BS 5534 : Part 1 : 1978: *Slating and Tiling* advises that trough valley tiles are not suitable for pitches below 22½ degrees. However, if the pitch is suitable, these look smart and match well with the main covering material.

Troughed valley tiles are made in concrete, and installation instructions are available from the manufacturers on request (*see* Fig 64). Preparatory groundwork will provide support using additional timbers, and procedures are detailed in manufacturers' literature. Preparation also involves making a cut-out on the upstand of the fascia board at the bottom of the valley. The lowest trough unit should be temporarily offered up so that a V notch can be marked on its lower edge to align with the angle of the eaves. When the notch has been cut with a Carborundum wheel machine, the trough tile can now be laid in position (*see* Fig 64).

The head valley tile should not be reduced in size and although minimum head lap for each trough unit is 100mm (4in), this may need to be increased in the first few courses to allow units to fit the given length of valley. With head and tail troughs positioned, the re-

Photo 71a Valley trough tiles. Commencing at the bottom, valley trough units are placed between battens with a lap of 100mm (4in) between each unit

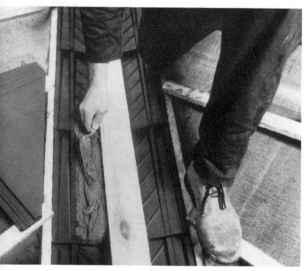

Photo 71b A length of timber, 100 × 50mm (4 × 2in) laid temporarily down the valley provides a guide for the placement of bedding mortar

Photo 71c The temporary batten facilitates accurate scribing in readiness for cutting tiles which border the valley

maining units should be laid between the enclosing side battens. The procedure is shown clearly in the sequence of photographs (see Photos 71a-71c).

If an open metal valley is preferred, the timber groundwork requires particular attention since it has to provide total support to the lining material, and creates the shape of the valley (see Photo 72a and Fig 42, page 80). Lead is one of the most common lining materials, and its various uses are discussed in the following chapter. A correctly constructed lead-lined valley performs well (see Photo 72b), but prefabricated valley gutters in rigid plastic have become increasingly popular. These are moulded to shape as shown in Photos 73a and 73b, and are much easier for an amateur to install. Support battens are needed on either side of the plastic extrusion, together with a central support running down the full length of the valley (see Fig 42, page 80). Lengths must not exceed 3m (9ft 10in), and laps should be at least 150mm (6in). After drilling the side lipping at metre (3¼ft) intervals, the extrusion is then nailed to the side battens.

Irrespective of the type of valley used, completing the installation follows similar procedures. If the roof pitch is less than 30 degrees, the finished valley gutter should be 125mm (5in) wide; pitches over 30 degrees can be served by a 100mm (4in) valley gutter. Tiles must be cut to align with the finished channel, and bedded on mortar. To do this, you should place a temporary length of timber, of appropriate width (100 or 125mm/ 4 or 5in), and 50mm (2in) deep, down the centre of the valley. This provides a straight edge, against which the side tiles can be aligned and marked. After cutting, they are laid on a mortar bed and placed hard up to the sides of the timber, as shown in Photo 71c. When this is removed you must smooth the face of the mortar with a trowel to leave clean sides.

Photo 72a On a slate reroofing project, an open valley is prepared for lining. Sarking felt has been laid over the valley boarding, and side battens added

Photo 72b The lead sheeting, in sections no greater than 1.5m, have been laid in the valley with overlaps of no less than 150mm

Abutments

Features like chimney stacks, dormer windows and parapet walls cause breaks in the continuity of roof coverings. Tiles will need cutting and you should endeavour to take them as close to the abutments as possible. However, a weathering strip is required as well, and flashing work is described in Chapter 7. A cement mortar fillet placed around the abutment is not a satisfactory answer. Differential expansion causes this to crack, and if the joint with the abutment severs, rainwater soon finds its way into the roof.

Special features

When constructing a roof with profiled tiles, several special components can be fitted. For instance, Redland manufacture transluscent-reinforced uPVC rooflight units to match all the profiled tiles in their range. These must be fixed with a piece of clear polythene substituted in place of the sarking felt, and no more than two units can be used in a panel. If positioned high up and in the vicinity of a loft access hatch, the benefits are noteworthy. Other 'specials' include Thruvent tiles with ventilation outlets. These ventilate the roof space, a useful point to note if you construct a mono-pitch roof for which ventilated ridge units may not be obtainable.

Photo 73a An off-cut showing the overall width of the unit and its shaped construction to fit against the batten groundwork

Photo 73b The plastic valley in situ showing the continuous guttering which is attractively suited to profiled tiles

Photo 74 An alternative to a lead slate flashing for weatherproofing pipes which penetrate roof coverings is to use the Redland 'RedVent' tile ventilation terminals. These units provide easy installation for the amateur builder

There are also vents for soil stacks as shown in Photo 74, and since these are moulded units there is full assurance of a weather-tight seal.

Dry-fix systems

A problem already highlighted is the vulnerability of tiles at the ridge and verge. Indeed there is a certain irony that it is now possible to purchase coverings guaranteed to last at least one hundred years which are fixed using traditional methods that may deteriorate within a decade. Ridge tiles are often dislodged in strong winds, and large cracks in the verge bedding mortar are a common sight. This is one reason for the growing interest in 'dry-roofing systems' which use what is described as 'mechanical fixing'. Attachment by wires, clips, nails, screws and straps is much stronger than 'wet-fix' methods with mortar. Improved structural integrity is an important benefit, but there are other reasons for its popularity, particularly the ease of installation.

In traditional practice, the mortar for verge and ridge tiles must be mixed correctly and consistently. Placing the right quantities, forming sound joints, striking off the excess and being able to point up the detailing is something which some amateurs find difficult. These tasks are also weather-dependent, and if it rains before the mortar has set properly, near-finished work is damaged. In contrast, a dry-fix system can be left at any point, irrespective of weather, and resumed later. When time is crucial, it is even possible to carry on working during light showers. With the growing need for ventilation in roof spaces to combat condensation, it is also significant that dry-fix ridges often incorporate air vents. Many high-volume builders are starting to use dry-fix systems on housing-estate developments, and although builders in the south of the country still seem to favour wet-fix methods, in the north it is claimed that fifty per cent of all new roofs are dry-fixed. The materials are slightly more expensive, but improved performance, easy installation and predictably fewer post-completion problems have evoked keen interest. With regard to self-build projects, dry-fix systems are understandably popular.

Both Marley and Redland have developed excellent systems along similar lines. If necessary, the verge can be treated separately from the ridge, although if you dry fix one, you might as well dry fix the other. At the verge, the Marley system employs uPVC sections which interlock and enclose the tiles. The sequence of photographs shows how this installation is carried out (*see* Photos 75a-75e). Each section is fixed individually, commencing at the eaves, and held by nails driven through guide holes into the end or upper sides of the tile battens. Nailing into end grain is not conducive to good fixing, but this is overcome by supplying annular ring-shank nails. These feature gripping collars along their shaft. The nail heads become progressively hidden as the sections are joined in succession. On reaching the peak of the roof, an end cap encloses the ridge tile and interlocks with the two uppermost verge units. Different cap ends are manufactured to suit different ridge tiles, and a mono-ridge version is also available.

Redland have developed two systems, one of which uses uPVC sections. An alternative arrangement, however, uses 'cloaked' verge tiles as shown in Photo 76a. These are shaped so that the tile wraps over the sides of the verge in one continuous unit. Stainless-steel clips (shown in Photo 76b) attached to the cloaked tiles create the mechanical fixing. Builders uncertain about the long-term durability of uPVC may favour the cloaked verge units, although this needs to be planned at the roof design stage since the overall width of the roof from gable to gable extremities should represent a 300mm (12in) measure.

For the ridge, dry-fix systems are particularly easy to install, and are suited to most interlocking concrete tiles – both flat and profiled. In the Marley system, lengths of shaped uPVC section are fitted instead of a top tile batten, and these serve three main functions: firstly the sections act as a support for the nibs of tiles on the top course; secondly, they provide a point of location for the sides of the ridge tiles; and lastly they provide attachment points for the security straps which are subsequently clipped over the ridge tile joins (*see* Photos 77a-77c). It is important that you fix these sections equidistant astride the roof peak, and at a width to match

Photo 75a Marley Interlocking Dry Verge System.
The special stop-end eaves unit is offered up against
an eaves tile to check the fixing position

Photo 75b With the eaves tile temporarily removed,
the stop-end section is fixed into place using a wire
hook nailed to the top of the fascia board

Photo 75c The uppermost ends of the units are fixed
by annular ring shank nails whose ridged shank
grips effectively when driven into the tile battens

Photo 75d As the subsequent interlocking verge
units are slotted into position nail heads are progressively hidden

Photo 75e The Apex is finished with an interlocking
ridge end cap and several patterns are available.
These include an optional bedding tie if it is preferred to complete the ridge with mortar

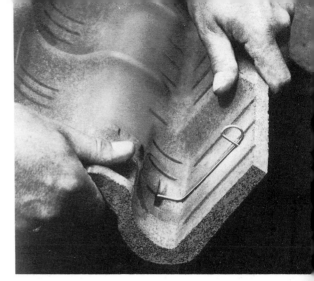

Photo 76a To eliminate the use of sand/cement mortar, Redland's cloaked verge tiles offer a neat finish and good resistance to wind lift. The maintenance-free unit is an interesting alternative to a barge board which needs periodic painting

Photo 76b Integral stainless steel clips on the underside of the Redland Cloaked Verge units give a mechanical fixing which is more reliable than a traditional mortar finish

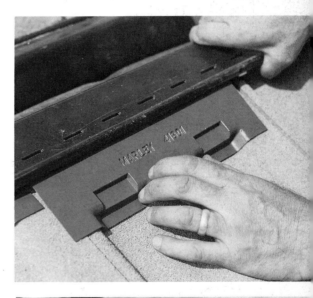

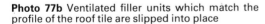

Photo 77a Marley Ventilated Dry Ridge System. Two uPVC batten sections are nailed either side of the ridge with the help of a setting out gauge which establishes correct spacing

Photo 77b Ventilated filler units which match the profile of the roof tile are slipped into place

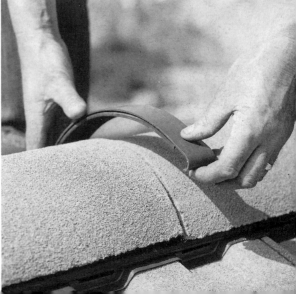

Photo 77c Ridge tiles are anchored to the uPVC batten sections with special union clips. When snapped into place, these locate between the ridge tiles and provide weatherproofing covering

Photo 78 An exploded view of the Redland Dry Ridge system. Note the metal clips which support a special ridge board to which the ridge tiles are nailed

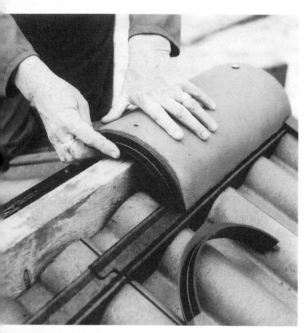

the base measurement of the ridge tiles. As shown in Photo 77a, this is achieved using Marley's clip-over setting gauge. The strap clips used to cover the ridge tiles provide a weather seal, and act in similar fashion to the straps clipped underneath the unions of uPVC eaves gutter systems.

The Redland Dry-Tech ridge system is different in so far as the top tile batten is retained. On a trussed roof, which lacks a ridge board, metal butterfly fixing straps must be nailed at the peak of each truss. The wings are bent upwards to enclose and act as a fixing for a ridge batten, whose size is determined by roof pitch and the tile profiles. Tables in the manufacturer's fixing guide indicate the required height of the batten, which will vary from 35 to 100mm (1½ to 4in). Ridge tiles are cradled in lengths of uPVC grooved section which is laid directly on the head of the top course, and these bear a troughed pattern where profiled tiles have been fitted. The Redland Dry-Tech system features sealing collars which fit underneath and between each join between the ridge tiles (see Photo 79a). Stainless-steel annular nails with neoprene washers and sleeves are used to fix the ridge tiles which are pre-drilled at manufacture (see Photo 79b). If you have to cut a tile, it will be necessary to form a new fixing hole. You will need an electric drill, a masonry bit and a measure of patience. Concrete is tough; progress will be slow and an excess of 'muscle' might 'blue' the bit or crack the tile.

Both the Redland and Marley dry-fix ridge systems are compatible with special fittings such as gas flue terminals and ridge ventilators. Some alteration in the groundwork has to be carried out, and in the case of the Redland system, part of the ridge tree will need to be removed, and a boxed unit constructed around the aperture to accommodate the flue adaptor. Means of attachment is

Photo 79a In the Redland Dry Ridge System, uPVC sealing pieces weatherproof the join between the ridge tiles

Photo 79b Aluminium alloy nails with integral washers are driven through preformed holes in the ridge tiles and into a ridge batten. This mechanical fixing combats wind uplift far more effectively than a mortar system

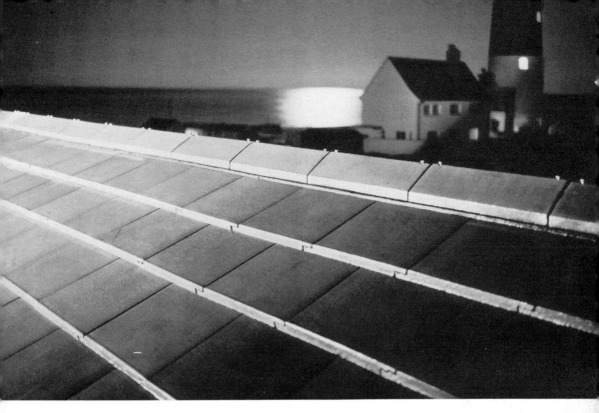

again mechanical – either by driving nails through fixing brackets into the edge of the ridge tree (Redland), or by strap clips (Marley).

Undoubtedly the dry-fix system has much to commend it, and there is no doubt that it will be developed with increasing vigour in the future. It is mentioned later that dry-fix systems are now available for slate roofs (*see* page 146), and although questions may be raised about the long-term durability of uPVC, the joy of the system is that sections can easily be dismantled, repaired and re-placed. Perhaps the claim that dry-fix is com-pletely maintenance-free awaits the test of time, but even the most entrenched tradi-tionalists are being converted. For the DIY builder, its arrival is particularly pleasing.

Double-lap roofing

The second method of fixing is used for flat covering materials which bear no interlock groove. This includes plain tiles (both con-crete and clay), quarried slate, synthetic slate and reconstituted stone. The absence of an interlock to provide overlap at the sides means that the entire covering has to incor-porate a form of bonding in adjacent courses. This is known as 'broken bond' and, to main-

Photo 80 Redland Dry Ridge units are available in different profiles. The Universal Angle ridge shown here is particularly suitable for their 'Stonewold' in-terlocking slates

tain the pattern, purpose-made tile-and-a-half units have to be fitted in alternate courses at the verge (*see* Fig 65). The installa-tion also involves double head lap, in which the heads of each unit are covered by two covering layers. In consequence, rain passing through each side join is arrested by the flat surface of the 'offset' tile below it. Although there is no weakness in the integrity of double-lap roofing, there is a greater total thickness of materials, greater weight, many more support battens and a good deal more work for the roofer. The principle of installa-tion is slightly different for tiles which have a moulded nib, as opposed to slates which rely solely on nail fixings. It is for this reason that tiles and slates are discussed separately.

Plain tiles

As shown in Fig 66 double lap means that the head of any tile will be covered by two tiles. This can often be seen at the verge where the mortar pointing emphasises the edge detail of the tile units. At the battens, the roof cover will attain a treble-tile thickness; when

133

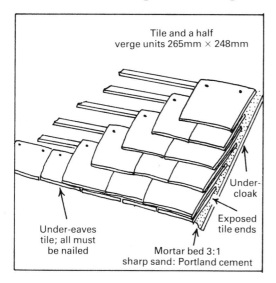

Tile and a half
verge units 265mm × 248mm

Under-
cloak

Exposed
tile ends

Under-eaves
tile; all must
be nailed

Mortar bed 3:1
sharp sand: Portland cement

Fig 65 In double lap roofing using plain tiles, tile-and-a-half units are needed at the verge in order to maintain the bond. An under-layer of shorter eaves tiles is also required, and these must *all* be nailed in place

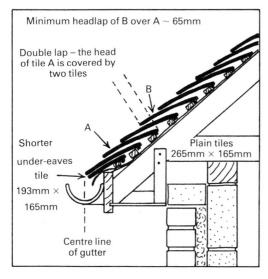

Minimum headlap of B over A ~ 65mm

Double lap – the head
of tile A is covered by
two tiles

B

A

Shorter

Plain tiles
265mm × 165mm

under-eaves

tile

193mm ×

165mm

Centre line
of gutter

Fig 66 Plain tiles are laid with 'double lap', which means that the head of each unit is covered by *two* tiles

the lap is measured, the centre tile in the sandwich is disregarded and the stated dimension relates to the overlap distance of the two outer tiles. Minimum head lap should not be less than 65mm (2½in), and in exposed situations, anything between 75 to 90mm (3 to 3½in) head lap may be required, although this depends on the overall size of the tile. As with interlocking tiles, an in-

134

crease in lap reduces pitch on the face of the unit. For this reason the head lap should never exceed one-third of the overall tile length. In the case of Marley concrete plain tiles, which measure 265 × 165mm (10½ × 6½in), maximum head lap to suit very exposed sites is specified as 75mm (3in) in their data sheets.

In respect of pitch angles, clay tiles in plain pattern should not be used on a rafter pitch less than 40 degrees; the lower limit for a concrete plain tile is a 30 degree pitch.

Gauge

Once again, there is an inter-relationship between lap and the spacing ie gauge of the battens. Limits are given in manufacturers' literature, and gauge can then be worked out as follows:

$$\text{Gauge} = \frac{\text{length of tile} - \text{lap}}{2}$$

As with single-lap tiling, batten gauge is the same as tile margin, ie the amount which is exposed.

To set out the roof, procedures are the same as for single-lap tiles, but there is an important difference at the eaves.

Eaves courses

In order to obtain a waterproof covering at the eaves, a double thickness of tiles is needed. This double layer must extend and overhang the fascia board to finish in line with the centre of the eaves gutter; 50mm (2in) is a typical projection. To achieve double-tile thickness, an under-layer of shorter 'eaves tiles' is required (*see* Fig 65). These are the first units to be laid, and the overall width of the eaves course must be ascertained using the same verge overhang prescribed for interlock tiles (*see* page 117). When centred relative to the verges, *every* eaves tile is nailed. To fix the shorter eaves tile, a special batten is required; its spacing from the batten above is worked out as:

$$\frac{\text{Length of a}}{\text{full tile}} - \frac{\text{length of the short}}{\text{eaves tile}} = \text{gauge}$$

With an eaves tile measuring 193mm (7⅝in) (Marley), this works out at 72mm (2⅞in).

Top tiles and batten

The same setting-out procedures apply to double-lap tiling as single lap, and the position of the top batten is established in the same way. A tile must be offered up temporarily, and positioned in accordance with the covering of the ridge tiles. A cover of 75mm (3in) should be regarded as minimum.

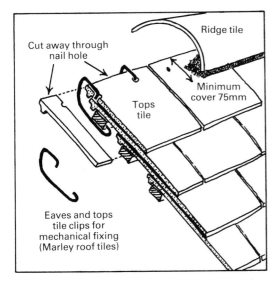

Fig 67 On a plain tile roof, the eaves and top tile units can be attached with a purpose-made tile clip. The clips shown in this cut-away diagram are available from Marley Roof Tiles

However, when plain tiles are used there is also need to include a covering course of shorter 'tops tiles' at the head to produce a double layer, and these are fixed in one of two ways. One method is to provide a normal batten fixing, although a thicker lath is needed. This should be twice the normal thickness, and most roofers fulfil this requirement by nailing one batten on top of the other (see Photo 70). An alternative method is to use a mechanical clip arrangement which holds the shorter tile in position on the back of the normal-length tile beneath it (see Fig 67). Detailing for this is provided in manufacturers' leaflets.

Verge finish

Much of the installation follows procedures described for single-lap tiles. However, an attractive feature is created if matching plain tiles are used for the undercloak instead of fibre-cement boarding (see Fig 68). In order to assist with the release of rainwater, the undercloak should be tilted slightly downwards to prevent droplets from being drawn back towards the gable. You should bed the verge tiles on mortar consisting of three parts sharp sand to one part Portland cement (by volume). They should be inclined slightly upwards so that discharging rainwater is directed towards the roof. Pointing up afterwards is a time-consuming task, but it produces a pleasing detailing which is unique to plain tiles.

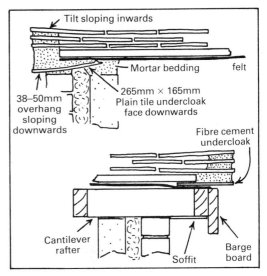

Fig 68 The verge on a plain tile roof can be mortar bedded on a fibre cement undercloak; a plain tile undercloak is a more attractive alternative but takes longer to construct. Note the need to direct rainwater away from the brickwork on the underside, and away from the edge on the upperside

Nailing

Plain tiles have no side interlocks to hold each other down and the nailing provision is important. All perimeter tiles should be nailed, and in respect of the remaining units, the revised British Standard recommendations (April 1985) are as follows:

> 35–60 degrees – all tiles in every fifth course
> Over 61 degrees – all tiles nailed

These suggestions assume normal weather exposure.

135

Photo 81a Plain tiles and slates are easily marked prior to making cuts to align with the rake of a hip or valley

Photo 81b Whereas a tile cropper or a carborundum disc machine is needed for accurate cutting, plain tiles can be roughly shaped with a pair of pincers

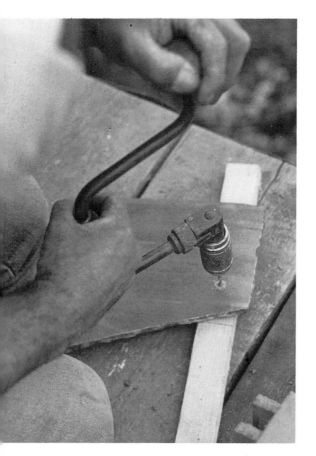

Photo 81c An electric drill and masonry bit is a preferred way to prepare nail holes in a concrete tiles, but where the mains power is unavailable, a brace and bit can be used

Tile cutting

On reaching abutments, roof windows or a hip ridge a number of tiles will need cutting. However, plain tiles are much easier to shape than profiled tiles. Cutting lines can be scratched with a nail, and a Carborundum disc machine will give a clean edge (Photo 81a). Alternatively, cut-outs can be made with pincers as shown in Photo 81b or with the help of hydraulic guillotines available from tool hire specialists. Holes can be formed using a masonry drill bit. (Photo 81c).

Hips and bonnets

On a hipped roof, roughly trimmed tiles at the end of each course are conveniently covered if you use segmental ridge units. However, on a plain-tiled hip roof, a neater answer uses cover units known as 'bonnets'. This is most attractive as shown in Photo 82, but it involves a higher level of workmanship, especially with regard to cutting. Bonnet tiles should align exactly with their adjoining courses, and must show clean butt joints. This calls for some precise cutting. Each bonnet tile is placed on a bed of mortar and secured with a 75mm (3in) nail to the hip-ridge tree. The lowest unit is usually filled with pieces of cut tile set in the mortar, which provides a neat finish and reduces the likelihood of mortar shrinkage. To the sensitive builder this finish is very attractive, though it is unlikely that the casual passer-by will notice. If time is limited, you are strongly advised to opt for segmental ridge tiles which are much easier and quicker to install.

Valley tiles

Just as bonnet tiles are an enjoyable feature not available for profiled roof coverings, so too is the purpose-made valley tile. The construction of a valley gutter was described in detail on pages 125-6. The principle of forming an open gutter can again be employed at intersections on roofs covered in plain tiles, but how much neater is the purpose-made valley tile alternative shown in Photo 83. Other attractive finishes are the traditional laced and swept valleys. However, these graceful features are not for the do-it-yourself roofer, and they involve a great deal of cutting and calculating.

To use purpose-made valley tiles, battens from adjoining roofs must be continued into the valley to meet in alignment (see Fig 69). They require support at this meeting point, and as with open valley gutters, supporting timber groundwork is required. In a traditional roof, a valley tree runs down the length of the roof intersection, and battens gain support from this central spine. On a trussed roof, however, timber valley boards are needed between the rafters, and battens thus gain support and a fixing point deep in the valley.

As with the bonnet tile, a purpose-made valley tile must butt directly against neighbouring plain tiles. Adjacent tiles thus have to be cut accurately to provide a neat finish (see Fig 69). But in spite of this task, the final appearance is worth the effort, and matching valley units suitably complement a plain tile covering.

Ornate ridge finishes

Plain-tiled roofs are sometimes capped with an ornate ridge tile. Such resort to ornamentation would be bizarre on a covering of profiled tiles, whereas on a plain-tile roof the effect is more appropriate. With increasing

Photo 82 On plain tile hipped roofs, cover units known as 'bonnets' provide an alternative finish. But the necessity for precise tile cutting at the end of each course and the need for horizontal alignment between bonnets and adjacent courses, calls for more skill than the installation of segmental ridge tiles

Photo 83 Purpose-made valley tiles suitably complement Marley 'Westwold' character roofing tiles

effort to reclaim and resell material from demolition work, original clay units are often used to good effect on new properties built in traditional styles. If you find it difficult to locate reclaimed materials, Red Bank of Measham, Staffordshire, is one of the few companies which specialise in crested ridge tiles and ornate scrolls and finials to grace the gable. A slotted ridge even enables you to insert your own choice of ornate finisher, with

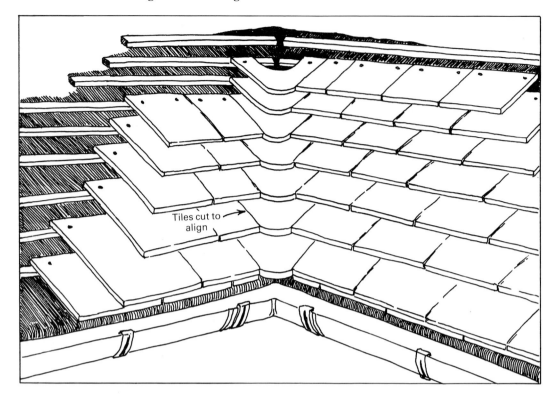

Tiles cut to align

Fig 69 Where plain tiles are used, purpose-made valley tiles give an attractive valley detailing. However, tiles located on either side of the valley units must be cut carefully and accurately

fleur-de-lis, holed motifs and arrow heads shown in Red Bank's catalogue (*see* Fig 62). In the right place, and on the right property, this ridge detailing can suitably enhance a roof whose covering is otherwise 'plain'.

Roofing with slates

Slate, whether natural or synthetic, is another double-lap roofing material. In many respects, slates are fixed like plain tiles, though the absence of nibs and reliance on nails is an important difference. When butted together, a join between adjacent slates must be situated over the central portion of the slate beneath to provide a waterproof barrier. This creates the familiar broken bond appearance, although slaters refer to this as 'half bond'. To understand this rainproof double layer, a simulation exercise can be done on the dining-room table with a pack of playing cards. The task is to cover part of the table 'slate fashion', using the lapping

principle. If the table-top is visible between the cards, your cover has failed. A further development is to use the nearest edge of the table to represent the eaves. This will show why a shorter slate is needed underneath the lowest course to create continuing double layering. In a similar way, the need for a shorter 'top slate' at the peak can be demonstrated with the playing cards.

You may find from this exercise that you can still provide cover by opening the vertical join between the cards. 'Open slating' as this is known provides better ventilation but less effective weatherproofing. It is often used on farm outbuildings, although in some parts of Northamptonshire it is seen on older houses. Rainwater finds its way under a slate by capillary action, and Fig 70 shows what is referred to as the 'angle of spread'. There is no need to describe the theory in detail except to say that the recommended pitch angles and laps are drawn up to ensure that there is no ingress of water into the roof space. Open slating is certainly more vulnerable in this respect.

In carrying out the card game, you will be employing the roofing principle used for

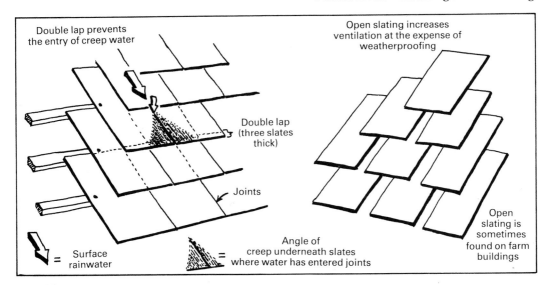

Double lap prevents the entry of creep water

Open slating increases ventilation at the expense of weatherproofing

Double lap (three slates thick)

Joints

Open slating is sometimes found on farm buildings

Angle of creep underneath slates where water has entered joints

= Surface rainwater

Welsh slates or their cement-fibre counterpart, in which units of regular size are used. However, some roofs are covered with slates of random sizes, and with the units installed in diminishing courses. The result is most agreeable, and can be employed with Welsh slate, Cornish slate or the sandstone slates typical of the Cotswolds. However, the expertise required to build a roof with random tiles and diminishing courses comes from years of apprenticeship; it is not a task for the amateur to attempt.

Notwithstanding this remark, two well-known manufacturers – Marley and Bradstone – *do* give clear guidance to the renovators of old properties who use their artificial 'stone' slates. For example, Bradstone reconstituted slates are made to look like the traditional covering, but bear code numbers cast on their underside. Installation instructions show how to set out diminishing batten widths, and explain a distribution system reminiscent of 'painting by numbers'. Documentation is particularly good, and on completion the traditional style and the fascimile stone slates are most convincing.

Tackling a roof with symmetrical quarried or cement-fibre slates is more straightforward, although it is advisable to commence on something small like a porch as shown in the photographs (*see* Photos 84a and 84b). However, there are important differences between quarried and synthetic material, and these are dealt with separately.

Fig 70 Capillary action induced by the flat surfaces of slates causes rainwater falling through the perpendicular joints to creep outwards. Double lapping and generous headlap prevents creep water from entering nail holes or passing over the heads of lower slates.

Open slating uses less units for a given area, but weather-proofing is reduced. The technique is unusual on domestic properties, but popular on farm outbuildings because it is cheaper and offers a higher level of ventilation

Working with Cornish or Welsh slate

Nailing
A special feature of slate is the fact that it can be nailed either at the head or near the centre of each unit. The different methods are shown in Fig 71, but centre nailing is by far the most popular form of fixing (see Photo 85).

The disadvantage with head nailing is the fact that it requires a greater number of slates to cover a given area, an increase in the overall weight and higher labour costs. The problem of breakage from wind lift is also greater due to the increase of leverage, particularly with long slates or when the roof pitch is low. Head nailing is not recommended on pitches below 35 degrees, although it is recommended for fixing short slates and heavy sandstone slates. Scottish practice still includes head nailing, but this is again only for heavy units which are smaller than 360 × 250mm (14 × 10in). Head nailing has been more popular in Scotland because northern quarries tended to yield much smaller slates.

Photo 84a A beginner's project – BEFORE
A small porch project which has had to be finished in sympathy with the slate roofs of the houses in the terrace

Photo 84b AFTER   The completed porch after two evenings work, with a slate covering in keeping with the rest of the property

Photo 85 Welsh slates from Penrhyn Quarry, near Bethesda, showing centre nailing and battens supporting the heads of each slate

It was once argued that an advantage of head nailing was the double cover over nail heads to produce treble thickness at these points. However, if slates are fixed in compliance with British Standard 5534 there is no advantage except for very small units.

Some advantages of centre nailing are a lighter covering, fewer slates, less labour and lower susceptibility to wind lift. It is also important to recognise that centre-nailed slates are easier to remove for replacement – a special bonus when repairing a roof. With centre nailing, a thin slate can also be used on pitches as low as $22\frac{1}{2}$ degrees. Certainly the majority of roofs built with large Welsh slates in the late nineteenth century were centre nailed. This is particularly desirable with long slates such as the 610mm (2ft) types, and also in positions exposed to prevailing winds. However, even with centre-nail fixing, the last course of slates at the ridge, and the under-slates at the eaves, have to be head nailed.

Present-day practice is to use a nail measuring from 32–63mm ($1\frac{1}{4}$–$2\frac{1}{2}$in), according to the combined thickness of the slate and its support batten. Copper,

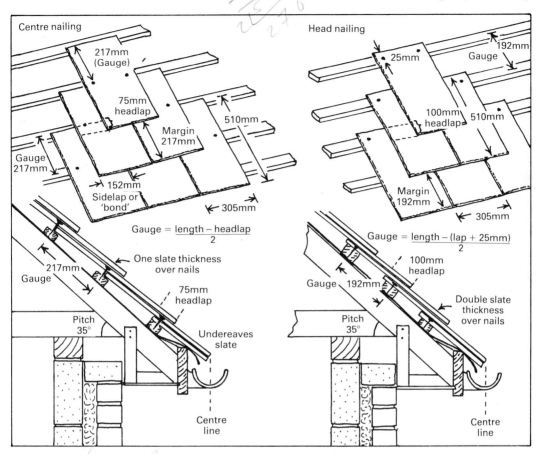

Centre nailing

217mm (Gauge)

75mm headlap

510mm

Margin 217mm

Gauge 217mm

152mm Sidelap or 'bond'

305mm

$$\text{Gauge} = \frac{\text{length} - \text{headlap}}{2}$$

217mm Gauge

One slate thickness over nails

75mm headlap

Pitch 35°

Undereaves slate

Centre line

Head nailing

192mm

25mm

Gauge

100mm headlap

510mm

Margin 192mm

305mm

$$\text{Gauge} = \frac{\text{length} - (\text{lap} + 25\text{mm})}{2}$$

100mm headlap

Gauge 192mm

Double slate thickness over nails

Pitch 35°

Centre line

aluminium or zinc nails should be used, because of their resistance to weather and chemical action. You must not be tempted to use spare iron nails from the tool box, or galvanised nails, whose life is similarly short-lived. Using wrong nails leads to problems, particularly in the acidic atmosphere of industrial areas or in coastal towns. A rusting nail can swell and fracture a slate; its life will also be far shorter than the covering itself.

Slates can also be held with stainless-steel hooks, but although used by professional slaters this method of attachment does not receive recognition in the current British Standard specifications. However, it is an alternative which is fully approved in the trade, and hooked slates on a large housing estate in Bangor have already lasted successfully for twenty years. At the intersection with rafters a hook is used which is driven into the timber whereas elsewhere an alternative type can be clipped around the batten as shown in Fig 72.

Fig 71 As a precaution against wind lift damage, centre nailing is generally used for Welsh slates, especially long units on low pitched roofs. However, the top course of slates, and the under-eaves slates have to be head nailed. Roofs covered with thick, heavy, short slates are usually head nailed; with this type of covering, the likelihood of snapping across nail holes, as a result of wind lift, is less evident

Hooks can also be used for repair work and are obtainable in small quantities and in various lengths from Penrhyn Quarries direct. Working out the hook length for a roof already nailed is shown in Fig 73. However, for the amateur undertaking to cover a new roof, nailing is the most popular fixing method.

As regards the nailing position for head fixing, a pair of holes are formed 25mm (1in) from the top edge of the slate, and no closer than 30mm (1¼in) from the sides. In centre nailing, holes are measured from the head at a distance equivalent to the gauge of the battens. This ensures that the lower edge of a slate will cover (ie lap) the nail heads of the slate which is fixed below it. Thus, although the term is 'centre nailing', holes are not exactly half-way along the slate length. In making these holes, the traditional method of punching with the sharp spike of a slater's axe causes the underside of the hole to flake away. If the slate is then turned over, the

chipping around the hole, or 'spoiling', provides a recess for countersinking the heads of the nails (Photo 86b). There is no point in attempting to copy the craftsman, unless you have a slater's axe (Photo 86a), plenty of spare time and a large pile of old slates to practise with. An easier alternative is to use a 150mm

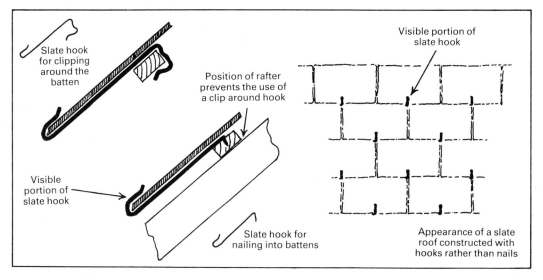

Fig 72 Two kinds of hook can be used for attaching natural slates and the type which clips around battens is particularly easy to install. However, this type cannot be fitted at points where the rafters and battens intersect and an alternative version must be used which is nailed into place

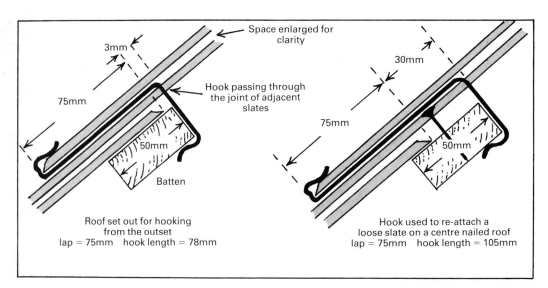

Fig 73 Short stainless steel hooks are used on new work, and their dimension and method of attachment is shown in the left hand drawing. On a roof where it is intended to hook slates from the outset, it will be noticed that a greater amount of the head lap is situated directly over the batten, and the hook in consequence is much shorter. On the right, a hook is used for a repair on a roof where slates were originally centre-nailed. But whereas the head lap of the slates is the same, its different juxtaposition relative to the batten calls for a longer hook. Calculating their respective overall measurements can be seen from these cross sectional diagrams

(6in) nail as a punch. Some beginners use an electric drill and a masonry bit to form the holes, but whereas this is a failsafe method, it is slow and fails to produce a recess for the nail head. In consequence, the slates do not bed down as closely together and are more likely to respond to wind lift. 'Kicking slates' is the slater's term for units which stick proud and are vulnerable in wind. In the south of England the term 'gapers' is more common.

Lap
As with plain tiles, slate head lap is determined by the degree of exposure, the size of the units and the pitch of the roof. Lap information is given in BS 5534 : Part 1 : 1978 (Tables 4 and 5), but this has been criticised from within the slate industry; Part 1 with all amendments is shortly to be reissued. New recommendations produced by slating specialists have been drawn up and revised information is given in Appendix 1, (Table 3).

Gauge
Once again, the distance between batten centres, ie the 'gauge', is related to lap and slate length. If slates are head nailed you will need to work out the maximum gauge with this formula:

$$\text{Gauge} = \frac{\text{Length of slate} - (\text{lap} + 25\text{mm}/1\text{in})}{2}$$

This is based on the assumption that nails are positioned 25mm (1in) from the head of the slate. In the case of centre-nailed slates, the maximum gauge is calculated as follows:

$$\text{Gauge} = \frac{\text{Length of slate} - \text{lap}}{2}$$

In setting out a roof, the same procedures obtain with slates as for tiles. With battens fixed at maximum spacing, it would be an unusual coincidence to find that they fitted into a ridge to eaves dimension exactly. In consequence, the gauge will be reduced to suit the roof, with the associated increase in head lap.

New or reclaimed slates
The information above is guidance if you construct a covering of new Welsh slates, Cornish slates or their synthetic counterpart.

Photo 86a Nail-holing practice using the spike of the slater's axe. By working from the underside of the slate, break-out around the hole on the face side provides a 'countersink' for the heads of fixing nails

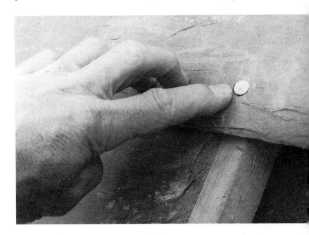

Photo 86b Break-out on the face provides a recess for the heads of the fixing nails. The beginner, who hasn't time to practise using a slater's axe, can achieve a similar effect using a 150mm (6in) nail as a hole punch

Quite often second-hand slates are used in order to save cost. These will bear fixing points from their previous installation, and you will be faced with the choice of either using the existing holes and working to the original gauge, or of punching new holes. Undoubtedly the addition of holes in close proximity to the originals reduces the strength of the slate. If the original nailing position offers a headlap gauge suitable for your roof, you might find it easier to use the original fixing holes.

It should be realised that slate is a natural product and there will always be minor differences in the thickness of units; it is most important to recognise this prior to installation.

Step-by-step installation for a centre-nailed slate roof

Disregarding special treatments which are dealt with separately, the main order of installation on a duo-pitch roof follows these steps:

1. Check the *actual* dimensions of slate before setting out the roof. You will find that different quarries produce slates with slight variation from the British Standard sizes.

2. Establish lap from table 1 in Appendix 1, and calculate batten gauge.

3. Hole most of the slates, remembering that distance from the head equals the gauge. While you do this, sort the slates into three groups according to thickness. The thickest will be used for the lower courses.

4. Felt the roof as described on pages 110–11.

5. Fix the top and bottom battens noting the positioning criteria described for tiling.

6. Establish how many slates fit across the width, noting that with a barge board no undercloak is needed. Check criteria already discussed with regard to tiles.

7. Coat the marking line with cement dye or ochre, and fix two temporary battens up the verges.

8. Establish where the perpendicular joins ('perps') will occur and, by stretching the line from eaves to ridge, 'ping' markings on to the felt.

9. Starting at the eaves double course, the under-eaves slates should be placed upside-down, ie bevel edge on the underside, to project 50mm (2in). You must cut shorter units for the under-eaves course whose length will be:

Gauge + lap + 25mm (1in)

The under-eaves slate is head-nailed, and usually shares the lowest batten.

10. Stack out the roof as shown in Photo 87 taking care to order the slates, thickest for the lower courses, thinnest at the top.

11. As shown in Photo 88, start from the bottom right (unless you are left handed), and build up diagonally.

Photo 88 Working diagonally across the roof, from eaves to ridge, is the usual way to build up slate coverings. Note how the perpendicular joints ('perps') have been aligned

12. Note that slates rattle if the nails are too loose, whereas tight nails cause the slates to break. If a slate is unwilling to bed flat and seems intent on wobbling, a trick of the trade is to cut off the top two corners (*see* Fig 74).

Photo 87 Slating a detached garage – an ideal project for the beginner. Slates are loaded out in the traditional manner, ready for use

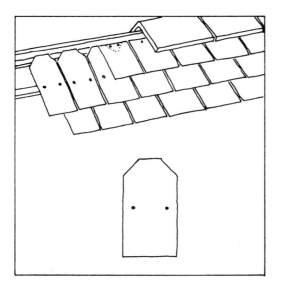

Fig 74 Centre nailed slates which tend to rattle rather than bed properly assume better stability if the top corners are removed. This technique can also facilitate top tile nailing by exposing the support batten

Beginners sometimes find the springiness of battens a hindrance when nailing, and there is nothing wrong in using a thicker batten to make the task easier. A trick used by slaters if battens are excessively springy is to drive in the nail with the wooden handle of a hammer, which reduces the bounce.

An example of a slate roof revealing several errors is shown in Photo 89. You will note the irregular thicknesses of the units which produce kicking slates, and suggest that no attempt was made to grade the slates. The bizarre alignment of the perps also shows that marking out was overlooked.

Cutting slates
The photographs (Photos 90a and 90b) show a builder cutting slates in the traditional manner for his self-build bungalow in Snowdonia. A sharp edge, a slater's axe, and a measure of confidence are needed. He decided to set up a break iron on a work horse, but smaller irons can be purchased from a roofing tool specialist. By placing the underside of the slate uppermost, the cut edge splinters to produce an attractive ragged bevel which gives character to natural slate. It is much easier to cut slates with a stone saw, but you will be denied the textured edging. A third method is to buy hand cutters from a roofing tool specialist, which resemble scissors with a

Photo 89 A slate roof which shows many errors. Failure to grade the thickness of slates so that thicker units could be used at the eaves, has led to random irregularities shown by the shadow in the horizontals. The perps (perpendicular joints) also show bizarre alignment. Accuracy can only be achieved with careful marking out

Fig 75 Cutting Welsh slate over a break iron, as shown in photographs 90a and 90b produces an attractive bevel edge. However, an easier alternative for the amateur roofer is to use a purpose-made guillotine or a parrot beak cutter – both of which are sold by roofing tool specialists

145

Photo 90a An amateur builder – photographed on the site of his self-build house near Snowdonia – demonstrates slate cutting. Note the static 'break-iron' – which he rescued from the derelict buildings of the former Llanberis quarry – and the slater's axe

Photo 90b With a slate resting firmly against the break-iron, sharp blows with the slater's axe trims a corner to suit the rake of a hip

Photo 91 A technique often adopted in North Wales uses ridge tiles to cloak verges on slate roofs. But marks on the gable wall are evidence of the problem of rainwater-creep under the ridge units

parrot's beak (*see* Fig 75). You will have to cut slates at abutments or to align with hips and valleys.

Eaves and verge details

To protect the gable, slates should project 50mm (2in) at the verge and be bedded on mortar (one part cement, three parts sand by volume) with a slight inward tilt. Where a roof finishes in close alignment with a brick gable instead of a timber barge board, this tilt is sometimes created with the addition of two or three 'creasing' slates bedded in the mortar. The tilt helps to steer rainwater away from the edge, but an attractive alternative is to angle the corners as shown in Fig 76. To preserve the broken bond in successive courses, every other verge unit should be a slate and a half wide; if these larger slates are difficult to obtain, half slates can be used instead. In Wales, slabs of slate (Photo 55) or ridge tiles (Photo 91) are often used to cloak the verge, but this leads to problems when rainwater creeps underneath the coverings (Photo 91).

Dry verge system

A development recently announced is a dry verge system for slate roofs. The 'Glidevale' verge illustrated (*see* Fig 77) comprises lengths of ABS/PVC and Polypropylene ex-

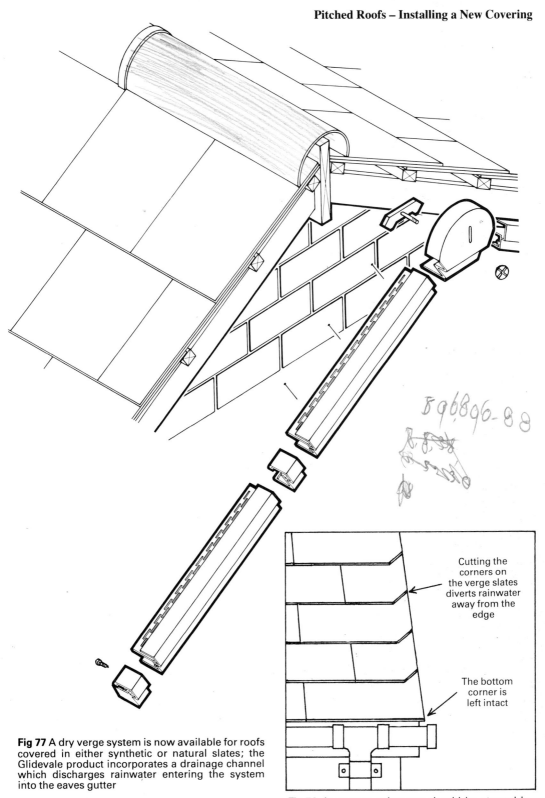

Fig 77 A dry verge system is now available for roofs covered in either synthetic or natural slates; the Glidevale product incorporates a drainage channel which discharges rainwater entering the system into the eaves gutter

Cutting the corners on the verge slates diverts rainwater away from the edge

The bottom corner is left intact

Fig 76 At verges, rainwater should be steered inwards rather than over the edge of a roof. When Welsh slates are used, angled corners fulfil this objective and create a pleasant detailing

Photo 92a Slabs of slate shaped to look like the more common roll finish created with sheet lead, as a hip covering are often seen in North Wales. Away from the areas of the slate quarries, other finishes are preferred

Photo 92b For the amateur, one of the easiest methods of finishing hips on slate roofs adopts the principles used for tiling. Hip tiles bedded in mortar cover the join, and examples in Staffordshire blue clays give a most appropriate colour match

Photo 92c A mitred hip in slate – an attractive alternative to the more familiar lead sheeting or hip tile cover methods. Weatherproofing is achieved by cutting and nailing 'lead soakers' around the hip at each course of slates

truded plastic which have to be nailed into place *before* the slates are installed. The interlocking arrangement makes provision for the problems of thermal expansion which occur with plastics. Sections are easily sawn to length, and the design is made so that any rainwater entering the units is discharged into the eaves gutter. Complementary end caps in a variety of patterns cover most shapes of ridge tile, and lock into the verge extrusion. The benefits of a dry-fix system have already been reported in the context of concrete tiles (see page 129). No doubt it will enjoy similar popularity when its virtues become more generally known. The fact that no special tools are required to effect a system offering better mechanical advantage than mortar will be of special interest to the DIY builder.

Ridge and hip treatments
Many slate roofs built in the early part of the century used dressed lead as a finishing cover for both hips and ridges. Firstly a rounded timber batten was nailed into place on the ridge tree or hip rafters, and then lengths of lead were folded around it and dressed to its curved profile. Leadwork is discussed in the next chapter, but it is worth mentioning here that on a slate roof, lead must be coated with patination oil (*see* Photo 103). Failure to do this results in conspicuous staining on the slates below. Certainly the alternative of installing a ridge tile is a much easier job for the

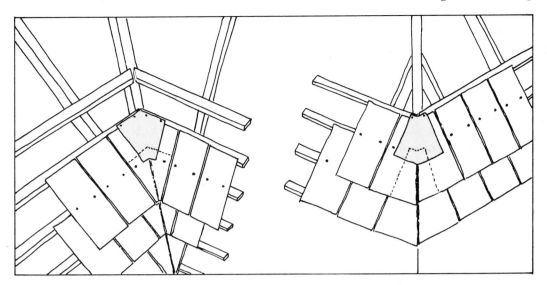

amateur, quicker to undertake and no less at-tractive. Whether you prefer matching grey ridge tiles or the contrast of a bright orange is a personal preference. Clay ridge tiles from John Caddick of Stoke-on-Trent have long been noted in the trade for providing a smart match or contrast with slate roofs (*see* Photo 92b). In any event, the points already made in respect of installing concrete ridge tiles are applicable to slates.

However, if you are patient, willing to cut slates with care, prepared to cut lead soakers, and eager to produce a high-class roof, you would eschew the above methods and opt for a mitred hip or valley (*see* Fig 78). This is highly attractive as Photo 92c shows, but it is not suitable on low pitches. In this instance the exposed cut edge should be sharp and not bevelled.

Ventilator units

The need to ventilate roofs is a subject dis-cussed in the next chapter. The Zamba aluminium slate-roof ventilator has been in-troduced, and fully complies with the re-quirements of British Standard 5250. Its Acrylic grey powder coating gives good colour match with both natural and synthetic slates, and it can be fitted as a replacement to a slate in around ten to fifteen minutes.

Synthetic slates

For the DIY roofer, cement-fibre slates are particularly easy to use. On account of their

Fig 78 Mitred valleys (left) and hips (right) are a most attractive feature on a slate roof. Waterproofing is achieved by using overlapping lead soakers at each course. These will be hidden when the mitred slates are nailed into place

light weight, they are suitable for reroofing where the integrity of roof timbers is less suited to the greater weight of concrete tiles. When dealing with hips, valleys or abut-ments, synthetic slates can be easily cut with-out special tools. By scoring a deep line on the face of the tile and flexing it across a sharp edge, a clean break is made. A fine-toothed handsaw can equally well be used if prefer-red. Units can be supplied pre-holed for centre nailing, and on account of their mate-rial, additional holes are easily formed with just a hand drill. An ordinary drill bit is sufficient, and a 4.5mm (3/16in) hole will accept the copper fixings. When manufactur-ers were consulted during the preparation of this manual, it became evident that plenty of back-up guidance on fixing procedures is available to both amateurs and professionals alike. 'Help-lines' offering technical advice represent another encouragement to use synthetic slates, which are certainly enjoying an increasing popularity. Manufacturers' clear instruction sheets would enable any practical person to carry out a roofing project with these slates, and only a précis is given here.

When comparing products, most manufac-turers offer both a nail-fixing system, and a

149

Photo 93 Ease of cutting, lightness, and ready availability makes synthetic slate an ideal covering for the self-build enthusiast to install. Manufacturers are also noted for their helpful literature describing fixing procedures

hooked-wire alternative. If the latter is preferred, slates are usually ordered without fixing holes. Both systems are different from the method used with quarried slates, and the variations should be noted.

Nail-hook fixing

Using one hook per slate, the sharp upper end is hammered into the centre of the batten. It is important to ensure that its position coincides with the break bond gap, that is the side join between the two slates in the course below. When the slate has been laid in place, the lower end of the hook is bent upwards to support the tail, making sure not to pinch it. At the margins of the roof (the verges or hip), nails are needed in addition to the hooks to ensure maximum resistance to wind lift.

The problem with this method of fixing is that the hooks' position is critical to achieve both vertical *and* horizontal alignment. For the experienced roofer, who is able to combine setting out accuracy with speed, a large roofing contract can be completed quicker with hook fixings; but for the amateur, the nail and disc rivet system is much to be preferred. Even among professionals, this is the more common choice, partly because nail hooking is not suitable on pitches below 25 degrees or in exposed areas. In addition to this limitation, it is worth mentioning that 'nail and disc fix' is the only method to comply with the British Standard Code of Practice for synthetic slates. You would be advised, therefore, to consider the second option.

Nail and disc rivet fixing

In respect of nailing, the synthetic slate is centre fixed exactly like its quarried slate equivalent, and you must make sure that the fixing is not driven too tightly against the face of the unit. The fact that slates are supplied pre-drilled is a time saver, and even if new holes are required you will find the material very easy to drill. Copper nails are supplied by the manufacturers, and you should specify 25mm (1in) examples for battens 19mm (3/4in) thick, or 30mm (11/4in) nails for 25mm (1in) battens. However, in contrast with natural slate, the cement-fibre version is lighter and has less tensile strength. Hence the problem of wind-lift damage is greater, and to avoid breakages, a tail-fix system is also used. Fixing procedures are shown in Fig 79. The slates are laid with a 5mm (3/16in) gap at the sides, to leave space for a copper disc rivet.

In appearance, the rivet is not unlike a large-head drawing pin. With the head downwards, it is inserted between the edges of the *two* under slates so that the shank protrudes upwards. When the slate in the next course is offered up, the pre-drilled hole in the tail is merely located on the shank of the rivet. Finally, this is bent downwards, making sure not to hammer the soft copper stem too tightly against the slate. The result is not unpleasantly conspicuous, but on single-storey buildings, the detailing is visible from ground level. With the disc rivet system, each unit is held down at the tail by the two slates in the course beneath it. However, special provision is needed at the eaves.

150

Eaves detail

In terms of tile projection and the raised position of the fascia board, the normal provisions are made. The difference concerns the arrangement needed with the disc rivet system, in which two shortened under-eaves slates are required (*see* Fig 79). Effectively, this gives a treble thickness of slates at the eaves, and all three units should align over the guttering (*see* Fig 80). The lowest two battens must therefore be positioned to achieve this objective. If you omitted the lowest layering, there would be no means for supporting the head of the disc rivet during installation.

Special verge detailing

Several finishes can be carried out at the verge, including the conventional pointed mortar bedding. For this arrangement, you should prepare a 1:3 cement/sand mix. Slates should be bedded on the mortar, and later struck off cleanly to give a smooth surface.

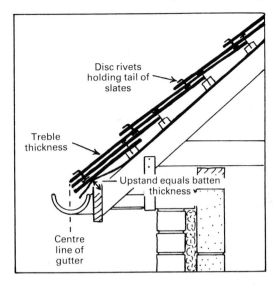

Fig 80 Using synthetic slates with disc rivet attachment at the tail, a treble thickness is attained at the eaves. The lowest eaves course is needed to support the head of the rivet

Fig 79 Procedures for installing synthetic slates using the nail and disc rivet fixing method

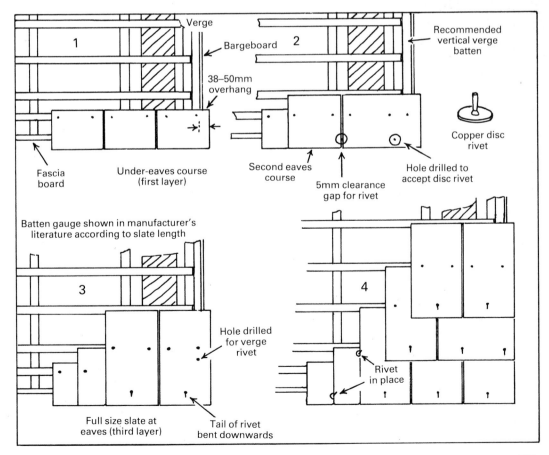

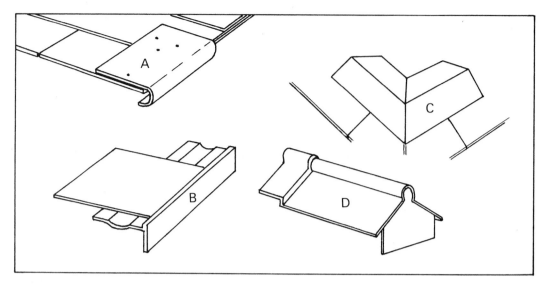

Fig 81 Ranges of synthetic slate include a variety of moulded units which produce a tidy finish on a roof. Some examples are:
A Wrap-over verge slate
B Moulded verge closure unit
C Angle ridge units
D Stopped end ridge tile in traditional roll-top style

As years pass, it is not unusual for this thin layer to crack and break away due to differential movement. It is on account of this weakness that one manufacturer offers a special wrap-over verge slate, or alternatively a moulded closure unit (*see* A and B in Fig 81). Both have similarities to the dry-fix systems used with roofing tiles. If you find it difficult to mix mortar consistently, or admit to a dislike of 'wet-trade' skills, these alternatives would be well worth your consideration.

Ridge alternatives
On account of the moulding process involved in the manufacture of synthetic slates, matching ridge tiles are included in most stock lists. Lead is thus unlikely to be used, but if you like the traditional appearance of the old roll-top ridge, the shape is copied in one of the option ridge units. You will also find angular ridges, hip ridges and even stopped-end units whose closure design obviates the need for a mortar fill at the end of the ridge (*see* Fig 81). Mechanical attachment is used, and socket/flange arrangement provides a means of interlock, which also dispenses with the need for mortar bedding or pointing. However, to supplement the integrity of the assemblies, mastics

are usually specified and supplied by the manufacturers. With regard to the attachment, brass screws, sealing washers and finishing caps are used. Installation is straightforward, and this represents a further swing towards the mechanical dry-fix philosophy.

Special fitments
In keeping with modern practice, means for ventilation can be built into the roof covering, and a variety of eaves intakes and ridge ventilators are included in most catalogues. Similarly gas ridge terminals are also manufactured in matching styles and colours.

Conclusion

The information contained in this chapter provides the main points about the installation of a roof covering. It has been addressed to the self-builder or amateur undertaking large-scale refurbishment work. Notwithstanding this 'readership', if you only need to carry out simple repairs it is helpful to understand the *methods* of roofing which have been described. However, the picture is incomplete because the instructions so far have been specific to particular types of covering material. Common to all is the need for a system of ventilation and for an additional weatherproofing cover referred to as 'flashing'. Since these topics are of importance, irrespective of whether your roof is covered in tiles or slates, they have been dealt with in a chapter on their own.

7 Ventilation, Insulation and Flashing

In this chapter, attention focuses on aspects of roofing which are important irrespective of the age of your home. The opening subject concerns an issue which has raised many worries throughout the building industry.

A problem identified – condensation

If attic insulation was the subject of the seventies, condensation in roof spaces could be described as the evil of the eighties. To be accurate, the condensation problem was recognised earlier, and a British Standard document published in 1975 concentrated particularly on the urgent need to control condensation in residential buildings. But it was not until 1981 that legislative measures were introduced in England and Wales, and these were set out in the Building (Second Amendment) Regulations 1981, under section 26 entitled 'F5 Specific requirements to limit condensation risks'. Whereas problems can occur throughout a building, the most worrying effect of high condensation levels relates to roof damage.

Without wishing to be frightening, the point must be stressed that condensation problems apply to *all* houses, regardless of age. Irrespective of whether your home is medieval or modern, the available facts clearly show that its long-term structure might be at risk. This assumes that you enjoy modern comforts like a washing machine, a bathroom, an airing cupboard, and live in a property in which modern materials have been incorporated in its structure. In view of these remarks, the owner of an older property might reasonably ask why it is suddenly deemed to be at risk when it has stood proudly upright for a century or more. Ironically, the problems are the outcome of modern life styles, and improvements which we have made to the fabric of our homes. With tongue in cheek, it can be readily admitted that you needn't read further if your home has no central heating, no additional insulation, draughty doors, ill-fitting windows, and a tin bath in an outhouse at the end of the garden. The condensation syndrome has unkindly emerged as a result of the quest for comforts. Compared to our ancestors, we produce more humid in-house conditions as a result of showers, dishwashers, airing cupboards and tumble driers. We also create a situation in which the air is able to *hold* more moisture as a result of central heating. Our modern residential envelope also reduces the chance for its dispersal by ventilation – modern houses are effectively sealed by efficient draught excluders on the doors, an absence of open fireplaces, and draught-free windows with sealing strips and double-glazing units. The unfortunate penalty of these familiar home comforts is the greater likelihood of problems from condensation – particularly in the roof space. There *are* cures, and you are encouraged to consider them. But firstly, it is appropriate to have a full understanding of this 'evil of the eighties'.

Causes of condensation

Concern about condensation in buildings prompted the publication of British Standard 5250, 1975, which is entitled *Code of Basic data for the design of buildings: the control of condensation in dwellings*. If you ordered this document through a local library, together with its 1983 amendments (AMD 4210), you would find it very readable, and pleasantly lacking in esoteric terminology. The cause and effect of condensation is concisely summed up in the Foreword as follows:

> In buildings condensation occurs when moisture produced by the occupants and their activities condenses on the internal surfaces of, or within, building elements. The dampness caused can be distressing to the occupants and can damage building construction as well as contents.

In modern properties kitchens, bathrooms and utility rooms are high moisture-producing rooms. Airing cupboards also contribute to the level of humidity. Equally significant is the fact that far greater amounts of water vapour can be held in *warm* air. However, if circumstances cause warm, moisture-laden air to cool, its vapour content is then released, which accounts for condensation. The temperature at which condensation commences to form is known as the 'dew point'. Cold surfaces can also create a cooling effect, which explains why cold pipes, toilet cisterns and metal window frames often carry condensate. This scientific fact has been evident for centuries, but the problems which have started to appear in roofs are recent.

Problems in the roof space

Ironically, problems usually begin in roof spaces as soon as loft insulation is installed. This remark is not intended to sow the seeds of doubt about the merits of loft insulation; the quest for better energy efficiency is unquestioned, and evidence to support these measures is well documented. For instance booklets addressed to householders from the Energy Efficiency Office underline the fact that if the loft of a heated house is uninsulated, heat loss through the roof is colossal. It is claimed that twenty-five per cent of heat loss may even take this escape route.

If a loft is lined with a ceiling-level insulant, heat is conserved more efficiently in the rooms below, but the loft space will be substantially colder in winter. At the same time, this insulation layer is not a barrier to the passage of vapour rising into the loft. Investigations conducted by the Building Research Establishment show that a large proportion of vapour finds its way into the roof space via the loft hatch. Almost as much enters around pipework, some vapour finds a way through ceiling light fittings, while joints or cracks in plasterboard also act as lines of weakness. But vapour can also permeate where air cannot, and some will pass through the ceiling itself unless it has been built using vapour-proof materials. The final pattern of events is self-evident. If moisture-laden air rises into the roof space, it is cooled, and the vapour content which had been successfully contained in the warmer rooms below is released as condensation.

Damaging effects of condensation in lofts

Condensation forms in two distinct ways. Surface condensate is easy to detect, and you should make sure that the cold surfaces of pipework and loft tanks are covered with an insulant. Instructions on how to carry out these important measures, outlined in the Energy Efficiency Office publications, are reproduced in Fig 82. In extremely cold conditions, you might also notice condensation forming on the underside of the sarking felt. This is more likely to occur on plasticised sarking materials rather than traditional bitumen products. At its worst, it is a sure sign of problems, and damage may occur if the loft is used for storage. For instance the binding of books may deteriorate, and ferrous metal objects like hand tools will rust quickly in this kind of environment. Leather goods may also become coated in a veneer of mould.

These problems are certainly inconvenient, but more serious is the damage which may be sustained by the roof structure. Surface moisture will quickly soak into roof timbers causing a further problem referred to as interstitial condensation. Vapour is a gas which can penetrate all but the densest of materials. Research would suggest that nearly all building materials can act as receivers, and if the vapour cools and condenses *within* components such as roof timbers, the eventual outcome is easy to appreciate. Moulds form, other fungal growth occurs and the roof could rot and collapse. In the event of surface condensation falling onto an insulant, its thermal conductivity may be impaired, and water reaching the ceiling material below would cause immediate damage. When considering the structural members of the roof, condensation wouldn't lead to deterioration overnight, but over a long term their life span would be significantly shortened. Fortunately you can take steps to avoid these problems, and manufacturers have been quick to develop products for new properties as well as components which can be fitted to the roofs of existing houses. As a DIY enthusiast you can easily take the appropriate action. However, there are two related matters to receive attention – insulation and ventilation. All home owners should pay close attention to these topics.

How to insulate water tanks and pipes in the loft.

One result of laying insulation in the loft is that the loft itself will get colder. Tanks and pipes within the loft will be liable to freeze up in cold weather, and for this reason insulation must be put round these items.

All cold water storage tanks in the loft, including central heating expansion tanks, and all water pipes in the loft, including overflows, must be insulated. Do NOT insulate UNDER the tanks unless they are situated well above the ceiling joists, since warmth from below should help prevent the tanks from freezing. The recommended thicknesses of at least 25mm (1in) for water tank insulation and at least 32mm (1¼in) for water pipe insulation should greatly reduce the risk of freezing in normal circumstances.

Insulation of cold water tank.

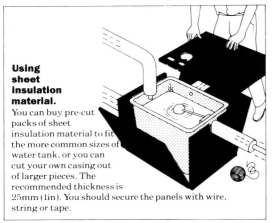

Using sheet insulation material.
You can buy pre-cut packs of sheet insulation material to fit the more common sizes of water tank, or you can cut your own casing out of larger pieces. The recommended thickness is 25mm (1in). You should secure the panels with wire, string or tape.

Using mineral- or glass-fibre mat material.
It is also possible to use mineral- or glass-fibre mat quilting secured with wire, tape or netting.

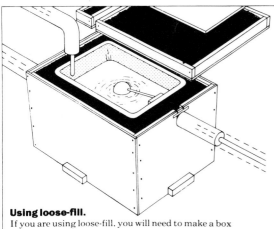

Using loose-fill.
If you are using loose-fill, you will need to make a box to contain it. This can be done with hardboard and timber.

Tank lid.
With both loose-fill and mat, you will have to make a lid to fit over the tank, which can be removed easily when you need to get to the ball-valve. You must make sure that no particles of the material can escape into the water. It can be a good idea to wrap polythene sheet around the lid and tape it up. Remember that provision must be made for any overflow from the expansion pipe.

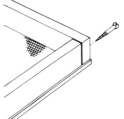

Cut two hardboard panels. Frame them. Join frame with screws. Pin hardboard to frame.

Cut two more panels. Frame similarly but set ends of frame an inch from each panel.

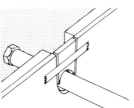

Cut out panels to slide over pipes. Assemble four sides pinning and screwing corner joints.

Make top from two panels with frame between. Fill cavity, taking care insulation doesn't get in tank.

Fig 82

155

Insulating roof spaces
Thermal insulation

Although the topics of this chapter apply equally to pitched- and flat-roof structures, it is the former which offers most opportunity for DIY enterprise. In new building work, minimum provision of a thermal insulant is mandatory. Details for domestic buildings were clearly specified in Part F of the Building Regulations 1976 (England and Wales), and prescriptive minima for houses and chalets north of the border are given in Section 1 of Part J in Building Standards (Scotland) Regulations, 1981. The high proportion of heat loss via the roof is due to the fact that hot air rises, and that most covering materials have a high emissivity level making them ineffective heat barriers. In recent years, the subject has received great publicity, and you can obtain free literature on all aspects of thermal insulation from the Energy Efficiency Office (*see* Appendix 2 for address).

When talking of thermal insulation, this is generally interpreted as the provision for controlling a *loss* of heat, but it should be added that heat *gain* is an equally important topic. On a hot summer day, heat gain in a loft space is considerable, and the role of insulation now works in reverse – namely to prevent high temperature levels creating discomfort in the rooms below. In either case, the need for an insulation barrier is abundantly clear. It is fortunate, however, that installing an insulation material in the loft space is a straightforward DIY job. The task requires no practical skills, although it isn't particularly pleasant. How easy it is in practical terms depends on whether the structure is a 'warm roof' or a 'cold roof'. It is important to understand these terms since they relate to the two different methods of insulating the roof.

Cold roof

The most popular way to minimise heat loss from the rooms below is to place an insulant over the entire attic floor area. Installing a layer of quilting such as fibreglass or mineral wool is one approach, although intricate roof shapes with variety in spacing between ceiling joists are more easily tackled by distributing special granules (eg exfoliated vermicu-

lite) over the whole attic floor area. This insulating strategy produces what is referred to as a 'cold roof'. The advantage of this approach is that the insulant is easy to lay in position, and the quantity of the material is kept to the minimum. On the other hand, the disadvantage of the cold roof is an increasing likelihood of pipes and tanks in the loft space freezing in cold weather.

Warm roof

If you convert a loft space into a habitable room or purpose-made store, a different insulating strategy is employed. In order to conserve the heat in this attic room, insulant must be placed between the *rafters* instead of between the ceiling joists. Warmth rising through the ceilings of the rooms below will thus contribute to the heating of the loft area, although raising the temperature of this increased living space would incur a bigger fuel bill. Moreover, this 'warm-roof' arrangement involves rather more constructional work than might be appreciated.

In respect of the insulating material, a greater quantity will be required to cover the inside of a pitched roof compared to covering the floor area. In consequence, the material cost will be considerably greater. Furthermore, producing an acceptable ceiling in the attic room means that the rafters need to be lined with a cladding material such as plasterboard. Before this is added, however, the insulant must be fixed in such a way that it is self-supporting and unable to slump down the rafters. You can do this by cutting thin battens to wedge between the rafters, thus holding it in place at frequent intervals. But there is also a need for a 'vapour barrier'. In the warm roof, unless prevented, vapour will pass beyond the insulation layer, reach the underside of a cool surface such as the sarking felt, and condense. One way to overcome formation of condensate on the felt is to construct a ducting to provide a throughput of air, between the sarking felt and the insulant. In practice, this is an involved construction, and the easier solution is to fit a vapour barrier. This is an airtight skin which prevents the passage of vapour. It should be situated on the *warm* side of the insulant, so that the barrier itself remains warm and does not offer a cool surface which would create condensation. Sheeting can be used, but a popular way to create a ˙barrier is to install purpose-made foil-backed or polythene-backed plasterboard. However, when you nail this into position, it is important to use sealing strips on the rafters where the board pieces are joined. Attention must also be given to the perimeters which need to include a form of sealing. All these measures must be taken to produce continuity in the vapour barrier, the effectiveness of which would otherwise be seriously impaired. If you come across the term 'vapour check', this refers to a partial barrier; foil-backed plasterboard fixed without sealed edges would be thus described. But, in the context of the warm roof, a continuous membrane situated on the warm side of the insulant is the only satisfactory answer.

Installing ceiling-level insulant
As a response to recent publicity, many loft-insulation contractors have appeared under the heading 'Insulation Installers' in the

Yellow Pages. If the professional is equipped with blowing machines, lofts can be insulated quickly; the hardest part is clearing the long-lost family heirlooms before the contractors arrive. But truthfully, self-help is worth considering since the work is notably straightforward. The accompanying illustrations (*see* Fig 83) show the procedures, and tell most of the story. Gloves are fully recommended, and a simple face mask is *highly* desirable. The latter is available from any specialist tool shop, and is unlikely to cost more than a pound or two. If you are working in an older property, the loft is likely to be very dirty and you might decide to hire an industrial vacuum cleaner for removing the dust. However, unless you later board the roof or add sarking felt, this cleanliness will be short-lived.

Photo 94 Quilted insulants are supplied in roll widths to conform with standard ceiling joist spacings. However, when narrower strips are required, the material can be easily cut with an old saw

Laying fibre roll

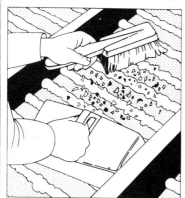

Get rid of layers of dust and any shavings left by builders. The surface should be clean before you begin to lay the fibre roll.

Put a piece of board across the joists and stand or kneel safely on it. Place the insulating material between the joists. Start at the eaves and unroll towards the centre of the loft. Make sure you leave a gap at the eaves for ventilation.

Cut roll at the centre of the loft and push lightly down between the joists. Start again from the opposite side of loft. Cut and butt ends together where lengths meet in the middle. Continue until whole area is covered. Do not insulate under the tank.

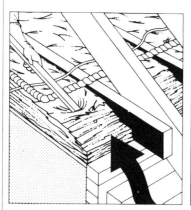

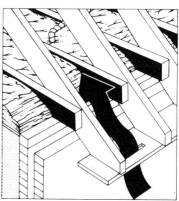

Cut fibre roll insulation to pass if possible under pipes and wires *being careful not to stretch or damage them.* Push into corners but remember not to seal the eaves completely.

Don't forget to fix a piece of insulating material on to the top of your loft trap-door. If mineral-fibre is used wrap it in polythene to avoid fall-out. Make sure that the trap-door fits properly, and draught-proof it with adhesive plastic foam. It may then need a catch or bolt to keep it tightly shut.

If the roof is boarded or felted under the tiles, so that you cannot see them, it is essential to make a gap or holes at the eaves to allow ventilation to your roof space. Seal any cracks around pipes and cables where they go through the ceiling.

General care

Laying loose-fill

Loose-fill insulation is simply poured from the bag between the joists, smoothed over with board cut to give the right depth when resting on joists. To reach into corners, nail board to broomstick. Most ceiling joists are 4 in deep, so filling level with the top gives good degree of insulation.

Fig 83

When dealing with ceiling joists at non-standard centres, granular fill has its advantages, but fibreglass can also be cut to fit. You will find that scissors are *not* the answer, but if you own an old handsaw, the rolled up quilting can be cut cleanly with a succession of light back and forth sawing movements (*see* Photo 94). Inevitably the loft hatch represents a line of weakness, and needs both a backing of insulant to reduce conductive heat loss and draught-proof strips to prevent warm air leakage. In practice this is quite difficult to carry out successfully with a standard timber hatch and lid. However, it is now possible to buy specially made hatches in glass-reinforced plastic which incorporate vapour seals. The Glidevale Access trap, for example, is moulded in such a way that the migration of moisture-laden air into the loft space is reduced to low levels. This attractive unit is fairly easy to install, and its self-coloured white plastic is one less item to paint when decorating the landing ceiling (*see* Photo 95). Needless to say, its contribution is only beneficial if all other routes for the passage of air *and* vapour into the loft have been closed as well.

Prior to the recognition of condensation problems, it was not unusual for the enthusiastic home owner to press insulation quilt into the sides of the roof, tucking it well down into the eaves. The message now is *don't!* In explaining the need to provide ventilation later on, you will see that on no account should the eaves unit be blocked. Since the objective is to prevent heat rising from the rooms below, there is obviously no point in extending the quilt beyond the limits of the ceiling. However, in a cavity-insulated property, it is recommended to tuck the insulant into the top of any cavity openings to prevent 'cold bridging' at wall-plate level. With this provision, the living space is thus enveloped completely.

When handling glass-fibre wool, you will probably suffer the unpleasant 'glass-fibre itch'. The purpose of gloves is soon evident, and arms should also be covered. But somehow the minute particles still seem to intrude and irritate. Some people are more susceptible to the problem than others, and the only cure is to wait twenty-four hours for the tingling sensation to disappear. It helps to rinse

Photo 95 To reduce the entry of vapour into the roof space, the Glidevale loft access trap, made in glass reinforced plastic, incorporates vapour seals around its perimeter

the arms in running water before applying soap and having a wash.

Local authority grant aid
The Government is acutely aware of energy wastage, and minimising heat losses has become a major issue in the design of *all* buildings. To encourage home owners to upgrade their properties, grand-aid schemes from local authorities have provided suitable inducements. But the ever-changing fortunes of local authority finances lead to ever-different support measures. Not long ago, grant aid wasn't forthcoming if there was already a thin veneer of insulant in the loft. Subsequently the ruling was changed, and a 'top-up' grant became available. Usually there are certain conditions imposed by the grant-aiding authority. For instance, one of their approved products will probably have to be used, and the work may have to be undertaken by an accredited contractor. Whatever constraints prevail, one point is always important to observe. *Do not* commence work or purchase materials until you have checked the authority's stipulations.

Controlling condensation

Research into the subject of roof-space condensation shows that low-pitched roofs are particularly at risk; the problem is also more acute if a roof is constructed without an eaves overhang. In seeking ways of controlling condensation, several strategies can be followed. For example, if the air temperature in the roof space equals that in the rooms below, moisture-laden air reaching the loft will not cool to the critical dew point. This 'warm-roof' approach would mean putting an insulant between the rafters as described earlier, but it would also mean higher fuel bills.

A wholly different strategy used for 'cold roofs' is to construct a ceiling through which *no* moisture-laden air is able to pass. This can be achieved if the entire ceiling area is provided with a vapour barrier on the warm side of an insulant, which is possible when constructing a new house or when carrying out a major rebuild. But in an existing property it is either impracticable or impossible. Making certain that there are no breaks in the vapour barrier, for example near ceiling roses, is far from easy. In consequence, one is left with a third and much preferred 'cold-roof' strategy, in which special provision is made to *ventilate* the loft space.

Air-flow alternatives
The rationale behind the strategy of ventilating a roof void is the fact that water vapour can be readily dispersed if there is a constant airflow. It is ironical that older roofs constructed without sarking felt or boarding are unlikely to suffer from condensation because they are naturally draughty. However, you musn't deduce *ipso facto* that the benefit of installing a secondary barrier is therefore held in doubt. On the contrary, it is important to have this barrier for reasons explained on page 109. No less ironical is the fact that bad carpentry work at the eaves will similarly ventilate the roof space. Gaping joints around soffits and fascias *do* create an unintended venting arrangement, but they also provide an ever-open door for birds, bees and vermin. Creating an effective airflow should be produced by design rather than by an accident of poor workmanship!

When considering ventilation in the duo-pitch roof, a cross-flow of air from *eaves to eaves* occurs if there are soffit ventilators and a wind blowing towards the fascias (*see* Fig 84). But a wind-dependent system is of limited use, coupled with the likelihood that a pocket of uncirculating air is likely to remain in the peak of the roof above the flow of air. Moreover, an eaves-only venting system is a non-starter in a mono-pitch roof or lean-to structure. Latest thoughts on roof ventilation favour an *eaves to ridge* passage of air. There are dozens of products which you can use to create this flow of air, and there are items to suit most roof coverings and different kinds of eaves. Some are intended for installation when a building is under construction, whereas other components can be fitted into existing properties.

Fig 84 Eaves to ridge ventilation is far more efficient than an eaves-only system which is wind-dependent

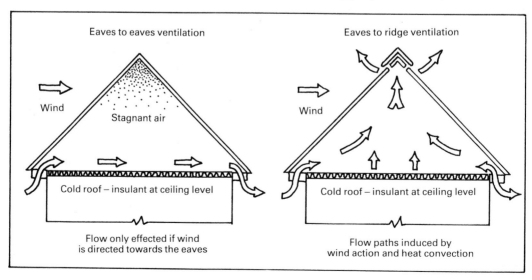

Eaves to eaves ventilation — Eaves to ridge ventilation

Wind — Stagnant air — Cold roof – insulant at ceiling level — Flow only effected if wind is directed towards the eaves

Wind — Cold roof – insulant at ceiling level — Flow paths induced by wind action and heat convection

Eaves venting
The provision of air vents in a roof covered with open-jointed materials, for example slates or tiles, is explained fully in the British Standard June 1979 amendment (AMD 3025) to BS 5250, 1975 (*Code of Basic data for the design of buildings: the control of condensation in dwellings,* clause 22.13). You will also find the chief features from this document embodied in the literature from manufacturers of ventilation components. Key points are as follows.

The provision is linked with roof pitch. For instance a roof steeper than 15 degrees should be constructed with an opening no less than 10mm (⅜in) wide, and running the entire length of both eaves. However, if the pitch is 15 degrees or less, a continuous opening of at least 25mm (1in) wide is needed. Unfortunately, it sometimes isn't possible to fit a ventilator extending from end to end along the eaves, in which case an alternative arrangement is needed. This is particularly evident on an existing building where to rebuild the eaves would be quite impracticable. In this situation, it is perfectly acceptable to fit separate ventilators spaced at regular intervals, *as long as* their total opening area is *no less* than the prescribed minimum achieved with a 10mm (⅜in) continuous strip. In other words, if a rectangular vent measuring 200 × 50mm (8 × 2in) is fitted into a soffit on a steeply pitched roof (ie in excess of 15 degrees), the area of ventilation would be the same as that provided by a ventilator 10mm (⅜in) wide, and extending for a run of 1000mm (39⅜in). Thus by positioning 200 × 50mm (8 × 2in) vents at spacings of 1000mm (1m[39⅜in]) from centre to centre, you will be achieving the same level of provision. But don't make the mistake of positioning them 1m (39⅜) apart; 'centre to centre' spacing means from the mid-point of one unit to the mid-point of the next.

Usually you will find this kind of detail in manufacturers' installation leaflets. Several products are described later in this chapter, and you will find no shortage of manufacturers who have launched systems designed to counter condensation. If you are tempted to cut costs by simply drilling holes in the soffit or fascia, you immediately declare 'open house' to birds, large insects and rodents. All proprietary ventilation systems feature some form of mesh, and you should regard this as *essential.*

When ventilators have been fitted, you must ensure that there is nothing inside the loft space which would act as a barrier to the air admitted at the eaves. It is for this reason that a quilt insulant like glass-fibre or rock wool must *not* be tucked down into the eaves. Indeed there should be *at least* a 25mm (1in) gap between the top of the insulant and the underside of the sarking felt or boarding, and, where possible, 50mm (2in) is preferred (*see* Fig 85).

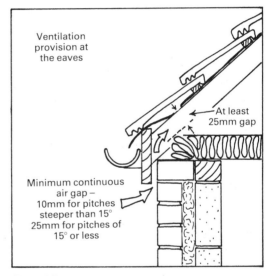

Ventilation provision at the eaves

At least 25mm gap

Minimum continuous air gap – 10mm for pitches steeper than 15° 25mm for pitches of 15° or less

Fig 85 Effective eaves ventilation relies on the size of the eaves opening, together with a gap of *at least* 25mm (1in) between the top of ceiling level insulant and the undersides of the sarking felt and boarding

Ridge ventilation
The same British Standards Institution document sets out details for ridge ventilation. The recommendation is again expressed in terms of a continuous ventilation gap, and 3mm (⅛in) is the minimum quoted. When building a new property, or completely replacing a ridge covering, a product like Marley's dry-fix ridge system is specially made with this provision. But, like the arrangement at the eaves, the area of ventilation is sometimes more conveniently situated at intervals. This is acceptable, provided that no pockets of air form outside the circulation routes, and as long as the overall vent surface equates with a 3mm (⅛in) continuous aper-

ture. It is interesting to note that products are now available to suit coverings in slate, clay tiles and concrete tiles – irrespective of size or profile.

As a final warning, it is important to state that ridge ventilation should never be installed without provision at the eaves. Failure to observe this leads to a suction effect in which moisture from the main body of the house is drawn upwards into the loft space.

Eaves ventilation in new roofs
When building a new house or undertaking a reroofing operation, a number of products are particularly suitable for eaves- and ridge-ventilation arrangements. For instance, the Redland RedVent eaves ventilator, shown in the accompanying photographs (Photos 96a and 96b), is very straightforward to install. The ventilator is made in black PVC, and suits roofs between 17½–70 degree pitch, where the rafters are spaced at 300–600mm (12–24in) centres. The unit is positioned on top of the rafters and eaves fascia *before* the felt and battens are added, and clout-head nails are used to secure it. Roofs with or without a soffit will accept the ventilator. A skirt along the fascia board is designed to overhang into the eaves gutter, and this is far better than the usual arrangement in which the roofing felt is draped over the guttering; projecting felt has a nasty habit of deteriorating along the gutter line. In addition to providing an unobtrusive 10mm (⅜in) continuous vent along the upper edge of the fascia, channelling moulded in the PVC ensures that air is admitted into the roof space along 25mm (1in) openings. The purpose of these moulded grooves is to ensure that layers of insulant are held away from the sarking felt at the low point of the roof. These units are simple to install, and you just have to add the sarking felt and battens afterwards.

A similar system is available from Marley using three main components – a strip ventilator, the underfelt support and channel ducting. Like the Redland product, these venting arrangements are intended to complement their various covering materials. However, if you roof or reroof with synthetic slate, several manufacturers provide similar systems to suit their product. For example, the range of TAC Duracem fibre-cement slates is supported by purpose-built eaves ventilator units; these are made in stainless-steel mesh for attachment on the upper edge of the fascia board.

An alternative to continuous ventilators fitted to *fascias* is the installation of a continuous grill fixed to *soffits*. These units, available from several independent manufacturers, are nailed to the lower part of the fascia board. Instead of the soffit extending to meet the fascia, it is cut to a reduced size and clipped into a housing provided in the ventilator.

Photo 96a To ensure that the passage of air entering via eaves ventilators isn't blocked by loft insulant, air channelling units are required. Redland 'Red-Vent' units, made in uPVC, are cut to suit rafter spacing, and then nailed into position

Photo 96b Once fixed, Redland RedVent ventilator channelling is covered with sarking felt in the normal way

Photo 97 Ventilation units can be fitted on top of fascia boards. But this arrangement is only practicable when reroofing or building new property

An example of this system is the Soffitex eaves ventilator which is made in galvanised-coated steel, and finished in brown, white and black to suit most requirements. A similar system from Glidevale is made in ultra-violet-resistant PVC. This is available in 2440mm (8ft) lengths, and the clipping groove accepts soffit board from 6mm (¼in) to 10mm (³⁄₈in) in thickness. Like the previous systems, these units are easy to fit when a new property is constructed; they are less easy to fit to existing buildings.

There is no doubt that the designers of roofing materials have responded quickly to the problems of condensation. Whether you tackle a roofing project yourself or enlist the services of a roofing contractor, look at the products available and make sure that the appropriate provision is made.

Photo 98 The Glidevale 'spring wing' ventilator is ideal for installing in the soffit of existing property. Once a cut-out has been made, the unit is held in place by 'winged' side clips

Eaves ventilation in existing roofs
If you want to take the all-important step to ventilate the roof space in an existing building, there are again plenty of products available. However, the idea of fixing a continuous opening is not practicable because its installation would involve too much reconstruction at the eaves. In preference, the manufacturers offer large-dimension grills which you have to fit at intervals. Usually these are installed in the soffit, and all you have to do is to form a cut-out. If room permits, a power jigsaw is ideal for the job. However, if obstructions affect the swing of the machine, a scrolling model (eg Black and Decker or Skil scrolling jigsaws) or a machine in which the blade can be fixed sideways (eg Peugeot 10TS) would cope best. If you don't own a jigsaw, an inexpensive pad saw – sometimes known as a keyhole saw – allows you to make the cut-outs by hand. Once the opening has been made, the rectangular ventilators are either clipped or screwed into place. There is no need for special fixings (*see* Photo 98).

As an alternative to rectangular cut-outs, you might prefer to install circular units like Glidevale 'Twist and Lock' ventilators. These provide openings equivalent to a 10mm (³⁄₈in) continuous vent when fitted at 200mm (8in) centres. A special holesaw for use with an electric drill is available from the manufacturer, and a neat feature of the unit is the fact that it is simply screwed into place in the circular cut-out.

Fitting ventilators is usually quite easy, although the different materials used for soffits may require special treatment. External-grade plywood is the easiest to cut; if tongued and grooved matchboarding is fitted, a little more care is needed when making the installation. But the most important note of caution concerns asbestos-cement soffits. Without needing to elaborate on the health hazard of asbestos-based materials, the point must be stressed that a proper mask should be worn whenever you are likely to create dust. In acknowledgement of the uneasiness about asbestos-formulated building products, you might even consider the possibility of replacing them with Ryton's ventilated soffit board. This is made in 6 and 9mm (¼ and ³⁄₈in) exterior water-bonded plywood, with integral vents offering a 10mm (³⁄₈in) gap for roof

pitches above 15 degrees, and a 25mm (1in) gap for roofs of 15 degrees or less. No special fittings are needed, and the ply is nailed to a batten fixed to the face brickwork, and a batten which you would need to fix to the back of the fascia board. If the board had originally been grooved to accept the soffit, this would be used in preference to a batten to house the material.

These measures are straightforward where a roof is constructed with eaves units. It was mentioned earlier that condensation problems are more severe in roofs where a soffit is omitted, and fascias fixed directly to the brickwork. In this situation, a remedy is less easy, and there may be a good case for removing and refixing the fascia boards proud of the exterior brickwork, to leave a 10mm (3/8in) or 25mm (1in) continuous gap. Purpose-made mesh should be added to keep out insects. If this is considered impracticable, another strategy is to introduce ventilating tiles at low level and spaced at intervals in place of the existing tiles or slates (see Photo 99). This is the only strategy possible when the eaves of the roof rest directly on a decora-

Photo 99 Glidevale ventilators, installed at low level, illustrate an arrangement often adopted when the provision of openings at the eaves is difficult to install

tive projection of bricks instead of a timber finish. Glidevale, for example, can supply ventilator units in colours, profiles and textures to match any slate or tile. Manufacturers like Redland and Marley also include ventilation tiles which match the different roof-covering materials in their ranges.

Ridge ventilation in roofs new and old
It has been mentioned earlier that some of the latest ridge tile dry-fix systems incorporate a continuous ventilator which runs from end to end of the roof. This is a splendid provision if you are building a new roof, but in refurbishment work, and many new projects too, ventilator units spaced at intervals are more common. Ventilation at the peak can be provided in two ways – either by installing large ridge ventilators at intervals, or with purpose-made venting tiles or slates.

Purpose-made ventilator tiles have already been discussed in the context of eaves provi-

165

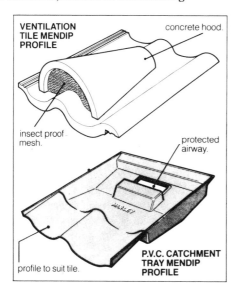

VENTILATION TILE MENDIP PROFILE

concrete hood.

insect proof mesh.

protected airway.

P.V.C. CATCHMENT TRAY MENDIP PROFILE

profile to suit tile.

Fig 86 Ventilation units from Marley feature an insect proof mesh and weatherproofing from a PVC catchment tray

land ridge ventilators, you will note that a continuous 10mm (⅜in) eaves opening or its equivalent at the eaves would require at least one ventilating ridge tile for every 2.8m (9ft 2¼in) along the ridge.

It is not a difficult task to remove an existing ridge tile and its bedding mortar in order to substitute a new ventilator unit, but on a traditional roof structure you would need to carry out some modification to the timbers. This would involve cutting away a section of the ridge tree and substituting a trimmer board on either side to make way for the vent duct projecting on the underside. Hence it is important that the positioning of the ventilator does not coincide with the rafters or trusses; it should fall between them. To get a correct location you will need several excursions into the loft to check and measure out the rafter positions relative to a gable wall or a hip. Assuming that the loft has been felted,

sion. If your roof is covered with Welsh slate, Zamba aluminium ventilators are specially made to blend with quarried materials by virtue of an Acrylic grey powder coating. It doesn't take many minutes to remove three slates with a slater's rip, cut out sections to accept the ventilator, and slide the slates and ventilator unit into place. Providing ventilation with special tile units is the only strategy employed on a mono-pitch roof (*see* Fig 86). The sharp, angular profile of the mono-pitch ridge tile does not lend itself to the inclusion of venting arrangement, as does the ridge tile on a duo-pitched roof.

A number of units are available for installation on a duo-pitch ridge. However, you mustn't confuse ridge ventilating units with ridge terminals for gas flues; they may look similar from ground level, but they perform a completely different function. Ventilating ridge units are available to match different ridge profiles, such as half round or angular, and their design ensures they are weatherproofed to withstand driving rain or snow (Photo 100). Calculating the number of ventilators needed is important, and you should consult manufacturers' literature for advice. The provision of 'outflow' at the peak of the roof bears relationship with the degree of 'inflow' at the eaves. For instance, if you look at the literature which describes Red-

Photo 100 The ridge ventilator units from Redland provide a smart answer to high level ventilation. These are available in various ridge tile profiles, and versions are available for both dry fix and mortar ridges

you will also need to cut an aperture in the sarking felt at the peak. Fixing a ventilating unit requires a mortar mix for bedding (*see* page 122), although some units incorporate a mechanical nail-fix arrangement. Exact procedures are well explained in manufacturers' literature.

Flat-roof deckings

Everything applicable to condensation problems in pitched roofs is again evident in the 'flat roof'. In this manual, we have limited our attention to flat roofs constructed in timber since these are most common in domestic premises. Concrete roofs, which are often used for industrial buildings or military establishments are not included. Usually the problems in a flat roof are more acute than in a pitched roof, and seeking an answer can be difficult. The fact that the void above the ceiling level is too small for access or close inspection is one immediate difficulty. Secondly, a flat-decked structure is a popular form of roofing over a home extension, in which case the roof will abut against a wall of the house. This immediately limits the opportunity to build an arrangement which facilitates a constant throughput of air. Moreover, it isn't unusual for an extension to provide additional bathroom, kitchen or utility room facilities – all moisture-producing situations. The final stumbling block is the fact that the main covering material is formed by a continuous barrier such as built-up layers of bitumen felt, which afford a nil level of ventilation. Not only is there a greater potential for the formation of condensation in flat roofs, but there is the added problem that the incidence of damp or mould formation *cannot* be observed without dismantling part of the structure. Like the pitched-roof structure, there are differences, however, between 'cold' and 'warm' roof constructions.

Cold-construction flat roofs

As mentioned earlier, a 'cold roof' has its insulant positioned at ceiling level so that the void does not draw heat from the room below. But whereas warm air is effectively contained by an insulating material, the passage of vapour is unchecked by fibre-glass or mineral-wool materials. Ventilation at the eaves is *essential* with a 'cold' flat roof, but at best this is only a token gesture if the joists run at right angles to the wall of the house. Joists separate the structure into channels, and on a lean-to building eaves ventilation alone does not induce a continuous flow of air. This was pointed out on page 49 and led to the amendments in British Standards recommendations (AMD 4210, March 1983, clause 22.9.2 for BS 5250, 1975). Hence in recent work, joists are often built to run parallel to the wall, thereby providing the opportunity to create eaves to eaves cross-flow ventilation. Since this usually involves bridging a bigger span, sturdier timbers will be required. But this is no help if you have an existing building where the joists run towards the house. In this situation it is essential to install a vapour barrier behind the ceiling material, on the 'warm' side of the insulant. This must be carefully constructed to provide a continuous membrane; a vapour 'check' as opposed to an unbroken barrier is not regarded as adequate. Various materials are available for this purpose, and heavy-grade polythene sheeting can be used, as long as any joins are folded together and nailed to a rafter. Another strategy is to buy an insulant which has a foil facing on one side, such as Gypglas by Gyproc Glass Fibre Insulation. Foil- or polythene-backed plasterboard is another popular way to form the vapour barrier, but joins *must* be properly sealed or the material will only be fifty per cent effective.

However, it *is* possible to create a continuous cross-flow if modifications are made to the roof. Creating a continuous run of ventilator grills along the soffit can be done with various proprietary systems, but this must be complemented by a series of roof vents situated near the wall of the main house. Ventilator openings topped with mushroom-like vents can be constructed, and if properly installed these will not be affected by the ingress of rainwater. A range of roof-level mushroom vents are available from Euroroof, but if you have any misgivings about cutting openings in roof covering material to fit these units, it is wise to seek the help of a roofing contractor.

Warm-construction flat roofs

A 'warm roof' in this context refers to a construction in which a vapour barrier is placed *above* the decking, followed by an insulant. The vapour barrier can be formed with a coated bituminous roofing felt (to comply with BS 747) which has been properly bonded at the joins to ensure a continuous seal. It has already been suggested that this kind of work is best undertaken by a roofing contractor. Hot bitumen poured and spread over the barrier adds to the integrity of the sealing, but this is not a task for the amateur unless he has equipment like a bitumen boiler. Alternative barriers such as aluminium foils sandwiched in bitumen are also available, and some types of roof deck can be obtained with an integral built-in vapour-proof membrane.

The insulation laid over the vapour barrier can take the form of a purpose-made insulating board. A sandwich which includes closed-cell polyurethane is a typical example. Particularly suitable for the DIY builder is a multi-functional product such as Coolag Purldek. This is a sandwich material supplied in 2400 × 1200mm (8 × 4ft) sheets which comprises three elements built into one. The top is formed by an 8m (3⁄8in) exterior roof-grade ply which forms the actual deck surface of the roof. This is bonded to urethane foam which is available in 25 and 50mm (1 and 2in) thicknesses. Finally a bonded layer of aluminium foil completes the hybrid, and offers a Class 1 surface spread of flame. Effectively, this product creates a vapour barrier, an insulated section and the top layer of decking – all in one product. It is fixed to rafters with clout nails and can then be topped by a conventional built-up felt covering. You should look closely at this material because it is an ideal product for the amateur to use.

Another way to install an insulant above a vapour barrier is to use a product known as Rubertherm manufactured by Ruberoid. This consists of block strips of polystyrene which have been bonded to a backing of bitumen felt. As shown in Photo 31 on page 50, Rubertherm is fixed to the vapour barrier using a cold bitumen adhesive applied from a gun. However, this should be regarded as a job appropriately assigned to a specialist contractor. Once laid, the final waterproofing layers of felt are added with conventional techniques such as the pour and roll method using hot bitumen.

Conclusion

Throughout this section the point has been stressed that the modern home-owner is able to enjoy an environment which offers all manner of benefits. Good central heating, methods of insulating the property and efficient draught excluders have helped to create creature comfort. But ensure that these are accompanied by measures which will eliminate the associated problem of condensation. You should regard insulation *and* ventilation as complementary tasks of modernisation.

Flashing and weatherproofing

The procedure for keeping out the weather still requires one further provision known as 'flashing'. Notwithstanding all the virtues of the coverings described, rain can still penetrate where the continuity of the surface is broken by chimney stacks, vent pipes or dormer windows. Moreover, a point of weakness occurs where the covering meets a vertical abutment such as a wall or parapet. A different strategy is required to provide weatherproofing at these junction points, and lead sheeting is one of the materials selected for this purpose. Lead is notably durable in all weathers, resistant to pollutants in modern industrial atmosphere, and easily worked into shape. It can even be used to produce decorative effects on roofs, although this is the preserve of the craftsman rather than the amateur.

Flashing and weatherproofing sheet and strip

Of necessity, this section is treated briefly, because there is enough material on the application of lead sheeting to fill a book. Indeed a specialist manual *is* available from the Lead Development Association (LDA), and this is regarded by architects and specifiers as *the* authoritative and definitive work of reference. A shortened version, entitled *Lead Sheet Flashings,* is also published by the Association in conjunction with the British Lead Manufacturers' Association. This is available free of charge from the LDA, and if you send a large A4 stamped addressed envelope to the address given in Appendix 2, you will receive all the guidance that you are likely to need. In recognition of this invaluable source, the aim here is not to repeat the A to Z of leadwork, but rather to describe the more common tasks which you are likely to come across. Moreover, this section doesn't focus exclusively on lead, but takes into account bitumen-backed metal foils. Arguably these are poor competitors, but in several situations this modern alternative is particularly useful. For example, in temporary repair work and short-term weatherproofing, it is a flashing material which holds special appeal for the DIY enthusiast. Its chief virtue is the fact that it is abundantly easy to install.

To return to the Lead Development Association, it is worth a reminder that in the building trade there are dozens of associations whose *raison d'être* is to promote certain materials. These trade associations represent sources of information for professionals such as architects, building surveyors or property developers. But some trade associations will also advise the amateur who is involved in self-build or refurbishment work. The LDA is no exception, and if you meet a problem not covered here, their major publication *Lead Sheet in Building – A Guide to Good Practice* (published by the LDA 1978, reprinted 1984, written by Charles Knight and Richard Murdoch) explains everything from constructing a simple flashing, to lead burning – which is welding in lead.

Flashing defined

Recognising the colloquial connotations assigned to the word 'flashing', it is appropriate to dispel any doubts and define the term in the context of building! In roofing, its definition is aptly summed up in *The Penguin Dictionary of Building* by John Scott as: 'A strip of impervious material usually flexible metal . . . that excludes water from the junction between a roof covering and another surface (usually vertical).'

If your roof abuts against a wall, a dormer window, or contains breaks for vent pipes, a flashing is needed for weatherproofing the junctions (*see* Photo 101). On some roofs you will notice that a fillet of mortar has been

used, for example around the base of a chimney stack, but this is not a successful means of weatherproofing. Different rates of expansion in brickwork, fillet and roof covering, coupled with structural movement, cause the mortar to crack or break away completely. The only satisfactory answer is to fix a flashing material. Whereas zinc, copper or asbestos-reinforced bitumen may be used, lead is most popular and assures a long-term performance. A sticky bitumen-backed foil is sometimes acceptable and easy to use, but in no respect is it a long-term substitute. If your roof is a simple shape, flashing with lead sheet undoubtedly falls within the DIY enthusiast's fund of skill. On the other hand, if your home sports a plethora of dormers and

Photo 101 The sides or 'cheeks' of a dormer window must be weatherproofed with lead flashing which overlaps the apron pieces installed on the front of the structure

169

Photo 102 When recovering a roof forming part of a terrace, the junction between new and old is best weatherproofed using lead. Covering the intersection with a hip tile is more popular, but differential movements between the coverings can break the mortar bonding

detailing like the grand roof of St Pancras railway station, you would wisely seek professional help. By tradition the plumber is the expert who undertakes leadwork on roofs on account of his trade's one-time association with lead for pipework. However, the traditional divisions of labour in the building industry have become blurred, and nowadays there are many roofing contractors who are equally skilled in leadwork.

Lead products

Lead sheet is available in several widths and six different thicknesses. A British Standard Code number is used to refer to thickness, although before metrication the material was identified by its weight in pounds per square foot. Modern packaging also employs a wrapper colour-coding system, and when you buy lead from the builders' merchant you should be aware of these details:

BS Code no	Thicknesses in mm	Packaging colour code
3	1.32	Green
4	1.80	Blue
5	2.24	Red
6	2.65	Black
7	3.15	White
8	3.55	Orange

In practice, lead sheet of Codes 3, 4 and 5 is what you are most likely to need for domestic roofs, as shown in the following recommendations:

To form 'soakers' for a slate or plain-tiled roof, use Code 3 lead.
Lining pitched valley gutters – Code 4 or 5
Apron flashing, eg at the head of a lean-to roof – Code 4
Weathering chimney stacks – Code 4, but Code 5 preferred for a shaped back gutter on the upper side of the brickwork.

In making these recommendations and when giving instructions later for several projects, it is assumed that you will be using lead sheet manufactured to British Standard 1178.

Fixings – nails and clips

Recognising the problems caused by metal mismatch, you should nail lead sheet with copper clout nails which have jagged shanks measuring at least 25mm (1in). Their large head holds the sheeting in place without any risk of puncture. To combat wind lift, any exposed edges of lead may also need to be held down and you can do this by making clips. Copper sheets at least 0.6mm (0.023in) thick can be used, but if the site is not exposed, you can cut out clipping strips from the lead itself. One end of the fastening can be hooked or nailed over a tiling batten, and examples are shown in the diagrams (Figs 91 and 98, pages 176 and 184). Finally, if you need to make fixings with screws, make sure that they are either brass or stainless steel.

Shortly after installation, rainwater flowing over lead flashing can stain tiles or slates with a film of lead carbonate – referred to as a 'patina stain'. This looks unsightly, and can

be avoided if you smear the lead with a special 'patination oil' (*see* Photo 103). Before this was available, plumbers often used to apply a smearing of linseed oil. Regrettably this final treatment of flashings is often ignored, and many smart roofs – especially those covered in darker materials like slate – are spoilt by white patina marks.

Tools

To cut lead, you will need a pair of tin snips, although a sharp knife can also be used for intricate work in thin sheeting. To shape the material – for example stretching it to seat around a profiled tile – special mallets are available. The technique is called 'bossing', and whereas traditional bossing tools are made from hardwoods like box or beech, toughened plastic versions are now sold as alternatives. In some texts you may come across the word 'dressing', which in plumbing also refers to shaping with bossing tools. This can lead to confusion because in the context

Photo 103 Applying patination oil on newly installed lead, reduces the likelihood of lead carbonate staining the slates or tiles

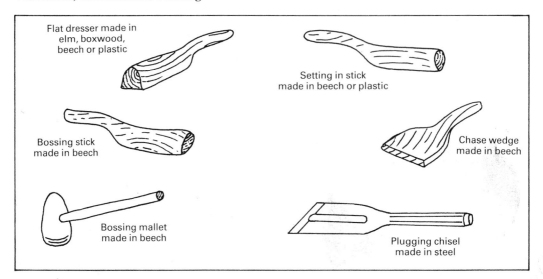

Flat dresser made in elm, boxwood, beech or plastic

Setting in stick made in beech or plastic

Bossing stick made in beech

Chase wedge made in beech

Bossing mallet made in beech

Plugging chisel made in steel

Fig 87 Specialist tools used in leadwork. The plugging chisel is used to cut a slot in a mortar course whereas the remaining tools are used for shaping and anchoring lead sheet. Although the use of plastic is acceptable for a few items like a setting in stick, the skilled tradesman has a clear preference for tools made in certain types of wood as indicated in the drawing

of slating, the term dressing always implies cutting. Lead is pleasantly malleable, and it may be tempting to try your hand at bossing with a hammer. But you run the risk of puncturing the sheet – not to mention cracking the slate or tile below. Traditional tools are shown in Fig 87, but for a one-off job, in which you don't want to spend money on a bossing mallet or dressing stick, form your own curved striker from an offcut of timber.

It is less easy to avoid spending a pound or two on a plugging chisel (*see* Fig 87). This is a sharp masonry chisel whose narrow shape is designed to remove mortar from between courses of brickwork. In some situations, you need to hire or buy an abrasive wheeled stone saw if slots have to be cut in concrete or rendering. These tools are necessary in making a provision to attach the sheet to vertical abutments.

Attaching lead sheet to vertical structures

Whether you construct what are referred to as aprons, stepped flashings, or cover flashings, the lead sheet has to be attached to a vertical surface. Where the structure is built in brick, you would cut out an attachment

slot with your plugging chisel. You then have to tuck the top edge of the lead into the brickwork to a depth of about 25mm (1in). Making a sound mechanical fixing is then achieved by rolling up small offcuts of lead – to about the size of hazel-nuts – and hammering them deep into the mortar course at approximately 400mm (16in) intervals. The lead ball jams itself into position to form an effective wedge. To complete the task, the course is finally pointed up with mortar to hide the wedges and to reinstate the integrity of the face brickwork. When you do this, dampen the brickwork so that the pointing mortar isn't sucked dry in a matter of seconds. The secret of successful pointing is to use a stiff mix of mortar which is worked from a small board like an artist's palette. You then cut lengths rather like cheese straws, scoop them on to the back edge of your pointing trowel, and press them accurately into the dampened slot.

If you have to attach lead to a concrete abutment, it is necessary to cut a receiving slot with a stone saw. The accompanying photograph (Photo 104) shows how this has been done on a concrete abutment bordering a slate roof. A similar procedure is sometimes used on rendered surfaces, but as the sequence of photographs (Photos 107a, b and c, page 178) indicates, it is not easy to carry this out neatly. A cleaner finish is achieved by removing a narrow band of rendering and fitting a length of galvanised-steel 'Renderstop'. This is used to sandwich the top 50mm (2in) of the lead to the abutment with stainless-steel screws, washers and wall plugs at intervals no greater than 450mm (18in). When the surface is made good, the replacement rendering will terminate neatly in the Renderstop to give an effective cover over the lead.

Notwithstanding the guidance above, weathering provision at junctions between a roof covering and a wall sometimes gives trouble. Quite often this is the result of poor pointing up afterwards, but for a complete summary of faults and cures you should consult technical information sheet 3.3 published by the LDA.

Simple operations

Apron flashing at the head of a lean-to roof

The point where a roof abuts against a wall requires a weatherproof covering. A fillet of mortar is not a successful way to provide this protection, and a flashing is the only long-term answer. Weatherproofing profiled sheet roofs is described separately on page 186, and the concern here is with provision of flashings on a tile or slate roof (*see* Fig 88). The structure might be part of a porch, garage or extension. You will need Code 4 lead which is sold in rolls of various widths.

When working out the width of the sheeting required, you must plan for an upstand of at least 75mm (3in), with a further 25mm (1in) to provide anchorage in the prepared slot or deepened mortar course. Calculating the extension of the apron on to the roof slope is dependent on pitch. A 220mm (8⅝in) cover is needed on a 20 degree pitch, 150mm (6in) on a 30 degree pitch, and 100mm (4in) on a 50 degree pitch. When completing the addition, don't be too disconcerted if you exceed the required width, even though the price of the product means that overestimating is a costly error. You can always trim the material down to size, using any offcuts to form wedges and anchor clips.

To take account of thermal movement, sections should be no longer than 1.5m (59in), and adjacent sections should be overlapped by no less than 100mm (4in). In effect, the expansion and contraction of lead is quite small, but take note of the recommendation. If the site is windy, make up

Photo 104 At the abutment on a slate roof, a cover strip, laid in lengths of 1.5m (59in) has been dressed over the upstand of the soakers, and tucked into a slot cut in the concrete parapet

Photo 105 An apron flashing made from Code 4 lead is a necessary weatherproofing cover at the head of the lean-to roof

173

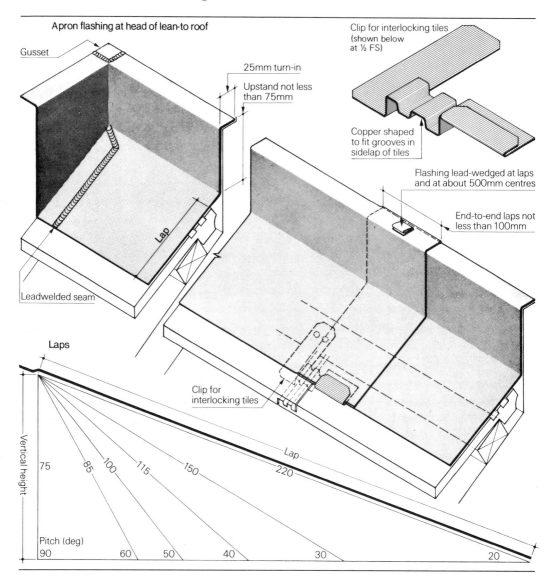

Apron flashing at head of lean-to roof

Gusset

Clip for interlocking tiles
(shown below
at ½ FS)

25mm turn-in

Upstand not less
than 75mm

Copper shaped
to fit grooves in
sidelap of tiles

Flashing lead-wedged at laps
and at about 500mm centres

End-to-end laps not
less than 100mm

Lap

Leadwelded seam

Laps

Clip for
interlocking tiles

Lap

Vertical height

75   85   100   115   150   220

Pitch (deg)
90   60   50   40   30   20

Fig 88 Detail of an apron flashing at the head of a lean-to roof. Finding out the required extension of the apron on to the roof is calculated by reference to the lower diagram

some copper clips to interlock with the tiles which must be bent over the leading edge of the flashing at every lap, and along the run of lead at centres no greater than 450mm (18in). Actual spacing will depend on the type of tiles used. Once positioned, the lead should be bossed to fit snugly around the tiles, and if your roof has a profiled covering, patient shaping work will achieve a neat finish. The malleability of lead never fails to please, and you will find this shaping a satisfying exercise.

Step and cover flashing for single-lap tiles
A roof covered with single-lap tiles featuring side interlocks is easier to deal with than a slate or plain-tile roof. Where the slope of the roof meets an abutment, a cover flashing is needed, and the saw tooth zigzag of lead is a familiar sight (Photo 106). For reasons described above, this is formed in lengths no greater than 1.5m (59in), and the surface cover over the tiles should be no less than 150mm (6in). Flat tiles in exposed situations – particularly laid at low pitch – are better served by 200mm (8in) coverage, with the added support of clips to combat the lifting effect of strong winds. If your roof is covered

Photo 106 A 'step flashing', with its characteristic saw-tooth pattern, is cut to suit the brickwork mortar courses at an abutment

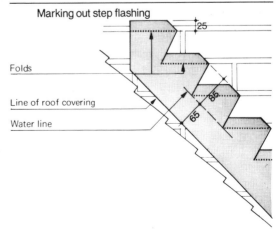

Marking out step flashing

Folds

Line of roof covering

Water line

25

65

85

Fig 89 The procedure for marking out a step flashing to cover plain tiles or slates installed with soakers. To cover single lap tiles, the same procedure is followed but the flashing needs to be 150mm (6in) wider in order to give a more substantial cover for dressing down over the roof units

with profiled units like pantiles, the lead should be of sufficient width to terminate well *below* the top curvature of the 'humps' on a down-turn. When calculating the roll width you need to buy, this measurement should be taken first. You should then add on 150mm (6in) for the wall cover.

Marking out the step flashing is shown in Fig 89 but an additional 150mm (6in) is required widthways to overlay the tiles, giving 300mm (12in) total width. An experienced plumber is able to offer up cut 1.5m (59in) lengths direct to the brickwork and mark off with speedy precision. Lead is too costly for error, and you will find it much easier to establish the shape required with paper patterns. An old roll of wallpaper is ideal, and a piece cut 150 × 1500mm (6 × 59in) can be offered up and pressed against the abutment to take a copy of courses and perpendicular joints from the brickwork. This is then cut and transferred to the 300mm (12in) wide

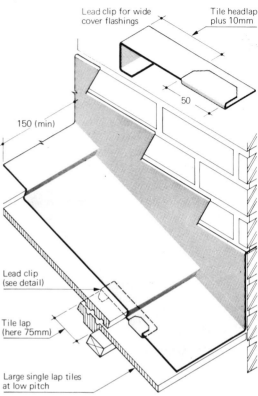

Fig 91 Combined step and cover flashing over single lap flat tiles needs the additional anchorage provided by lead clips

length of lead. A scaffold plank placed on trestles provides a useful temporary workbench while the lead is cut to shape.

Fixing the step flashing involves the same methods described for aprons; mortar courses must be deepened, lead wedges should be used to anchor the folded lugs, and pointing up with mortar completes the installation. Combined step and cover flashing for contoured tiles is shown in Fig 90; the finish on large flat tiles and support clips is shown in Fig 91.

Double-lap tiles and slates – soakers and cover flashings

Both plain tiles and slates require the additional provision of lead units described as 'soakers'. As shown in Fig 92, these are small units which are positioned beneath each slate and turned upwards at an abutment for 75mm (3in). They are made from Code 3 lead, and are cut 175mm (7in) wide, which in-

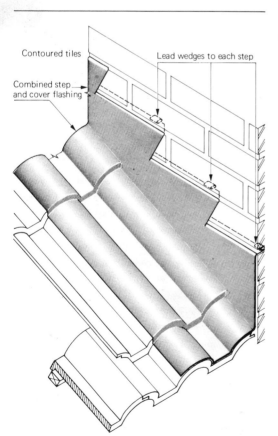

Fig 90 Detail of a combined step and cover flashing used for single lap tiles

cludes the 75mm (3in) for the upstand; length is calculated as follows:

Length of soaker = gauge + lap + 25mm (1in)

When positioned, each soaker is turned down 25mm (1in) over the head of each slate or tile so that the unit doesn't slip out of place. Making soakers is a simple operation, but purpose-made units are available to suit some coverings. For instance, a semi-rigid asbestos bitumen soaker is made by Redland

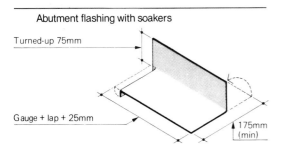

Abutment flashing with soakers

Turned-up 75mm

Gauge + lap + 25mm

175mm (min)

Soakers fixed in position

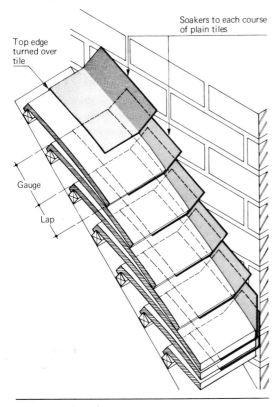

Soakers to each course of plain tiles

Top edge turned over tile

Gauge

Lap

Fig 92 Prior to the installation of a step and cover flashing, double lap tiles or slates must be weatherproofed with soakers which are placed beneath each unit and turned up alongside the abutment

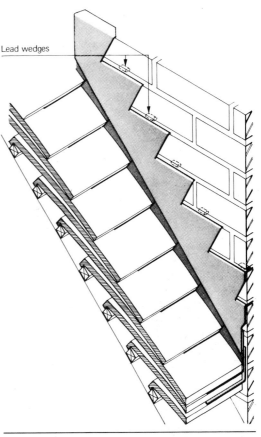

Lead wedges

Fig 93 With double-lap slates or tiles, soakers have to be overlaid at a brick abutment with a stepped cover flashing

for their 'Stonewold' interlocking slates. Its shape is similar to the lead soaker illustrated in Fig 92, but when installed it is visible on the surface of the slate; normally a soaker is hidden *underneath* the covering units.

Soakers are insufficient on their own, and to complete the weatherproofing they have to be overlaid at the abutment with a cover flashing (*see* Fig 93). On brickwork, this will be formed from 150mm (6in) lengths in the stepped configuration already described for single lap tiles. You make this up in the same manner but do *not* need to fold it over the face of the tile or slate. In consequence, you will only need material 150mm (6in) wide to form the stepped covering. If working against a rendered or concrete abutment, a strip attached to a continuous slot – without any stepping – is all that you require. This is shown in Photo 104, page 173.

Using cavity-tray products

A different way to create a stepped flashing involves individual cover pieces, manufactured in conjunction with moisture collectors known as cavity trays. These are shown in Photo 108. With a few exceptions, a cavity tray is normally installed when a property is first built. The object of fitting these units is in recognition that exposed face brickwork absorbs rainwater which can seep into unwanted places. A tray acts as a barrier as well as a collector of cavity condensation and residual moisture from the brickwork. This is subsequently discharged on to a roof via small weep holes. The function of a cavity tray is not dissimilar from a damp-proof course, except that it collects and disperses water which might percolate *downwards*. If the upper portion of an external wall is exposed to the weather, but its lower portion forms an internal wall for a room below, driving rain can travel down the permeable brickwork and damage the plasterwork. A typical situation occurs when a home extension is built against a wall which hitherto was exposed to the weather. The answer is to install cavity trays, and Photo 108 (opposite) shows a rather special type. The product is made in different shapes to suit various applications, but the interest in this section concerns the type intended for gable abutments.

The function of cavity trays is a wholly separate issue from roof flashing, but the fact that Cavity Trays Ltd of Yeovil are now producing a tray which doubles up as a flashing cover unit is worthy of mention. Here is a simple way for a builder to resolve two problems, with a particularly easy answer to stepped flashings, as shown in the photo.

Leadwork around stacks

Using lead to weatherproof the point where a vent pipe penetrates the roof is not a DIY

Photo 107a Installing lead flashings on a new plain tile roof. Individual 'soakers', cut from Code 3 lead, are hooked over each batten. Their upstand is then dressed close to the abutment using a bossing stick

Photo 107b A cover flashing is prepared in Code 4 lead for weatherproofing the soakers

Photo 107c The cover flashing is dressed to the shape of the tiles to complete the detailing

task since it involves lead burning. Moreover, there are several modern alternatives involving fabricated tiles which feature a rubberised collar to grip around the pipe. To suit some tiles, there are also special terminal pieces which are moulded complete with a uPVC tile. However, if you prefer to follow traditional practice, Cavity Trays Ltd operates a special fabrication and lead supply service. The arrangement involves what is known as a 'lead slate', which is designed to interlock with the coverings according to customer specification. The cover collar which will sleeve around the pipe is then lead burned to the lead slate or tile by the manufacturer.

Flashing around a chimney stack
When dealing with a chimney stack, Code 4 lead is recommended for sides, apron and back gutter. This assumes that no lead burning is required on the back gutter, in which case Code 5 lead would be needed. Where a stack is positioned astride a ridge, a back gutter is obviously not featured.

Photo 108 An alternative method of making step flashing can be followed when cavity trays have to be inserted in an abutment wall. The units from Cavity Trays Ltd, are made integrally with individual lead flashing pieces to create the step covering

The first flashing to fit is an apron piece, in accordance with procedures already described (*see* Fig 94). However, you must extend the apron 150–200mm (6–8in) either side of the stack, and the upstand must be dressed around the sides. Side flashings are fitted next, adopting the procedures detailed for gable abutments. An overlap is essential, and side pieces must be dressed to fit round both the upper and lower corners (*see* Fig 94). Effectively this means that there is a double thickness of lead where the side pieces overlap the apron.

Special care must be exercised when you form the back gutter since this will receive rainwater discharging from the upper part of the roof (*see* Fig 95). Two pieces are required – the main section for the gutter and a cover strip to attach to the rear of the stack. Dimen-

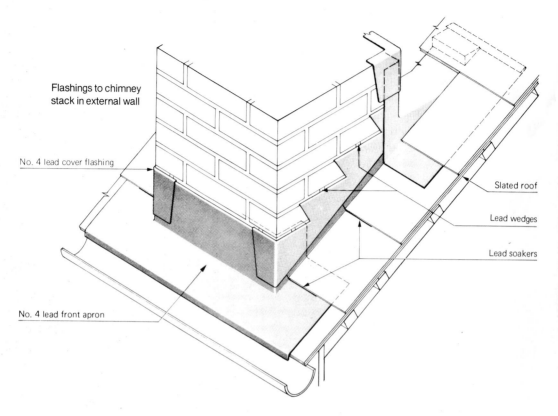

Flashings to chimney
stack in external wall

No. 4 lead cover flashing

Slated roof

Lead wedges

Lead soakers

No. 4 lead front apron

Fig 94 Detail of flashings to cover a chimney stack.
Work commences by fitting the front apron

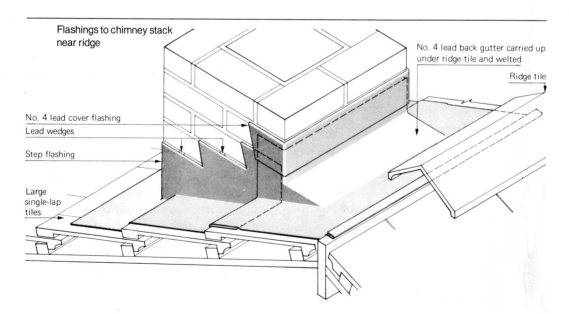

Flashings to chimney stack
near ridge

No. 4 lead back gutter carried up
under ridge tile and welted

Ridge tile

No. 4 lead cover flashing
Lead wedges

Step flashing

Large
single-lap
tiles

Fig 95 The back gutter on a chimney stack is formed
from two sections – the main back gutter section,
and a cover flashing fixed to the stack

180

Photo 109a Detailing on a chimney stack. The back gutter of a chimney stack must be prepared with care. A large piece of lead is required, though only part of the finished gutter will be visible

Photo 109b A bossing mallet is used to dress the back gutter around the sides of the stack; this covering piece is finished last

sions for the section required to form the gutter will be as follows:

Width–stack + at least 225mm (9in) on either side (more for a profiled tile)
Upstand at rear of stack – 100mm (4in)
Sole of gutter – at least 150mm (6in)
Extension piece for the roof slope – 225mm (9in)

The success of the gutter is dependent on sound groundwork, and the base should be built from timber and made structurally sound. The uppermost part of the section should terminate in a welt, which simply means that you fold back the lead, thereby preventing an ingress of rainwater. Forming the complete unit is an involved operation and the integrity of the gutter is wholly dependent on careful workmanship. Fig 95 illustrates the shape of the finished unit, and if you have any misgivings about tackling the job don't hesitate to enlist the help of a plumber. The method of weatherproofing a chimney stack at the ridge of a roof is shown in Fig 96.

Secret gutters

Occasionally you might come across a recessed and narrow gutter on either side of a chimney stack. This construction is nowadays discouraged because the narrow channel is all too easily blocked. If you insist on featuring this weatherproofing detail, consult the LDA manual for information about fabrication procedure.

Saddles

A saddle is appropriately named on account of its shape. A typical example is a ridge saddle, and this is shown in Fig 97 and Photo 110. If a ridged dormer window springs from the main roof, there is no way that either tiles or slates can interlock or overlay each other to produce a weatherproof finish. Lead produces the easy answer, and in effect you simply make a cover unit which bridges the junction of the dormer ridge with the slope of the roof. You must extend this well below the covering units as shown in Fig 97, and ensure that it is secured by dressing it over a batten.

181

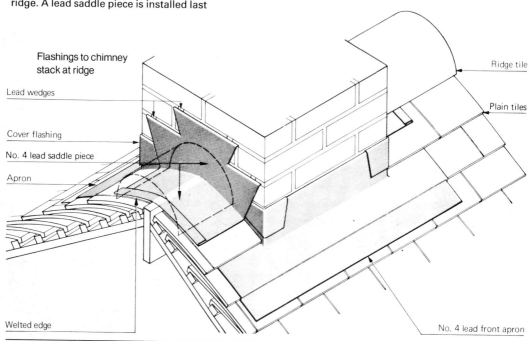

Fig 96 Details of flashing to a chimney stack at the ridge. A lead saddle piece is installed last

Flashings to chimney stack at ridge

Lead wedges

Cover flashing

No. 4 lead saddle piece

Apron

Ridge tile

Plain tiles

Welted edge

No. 4 lead front apron

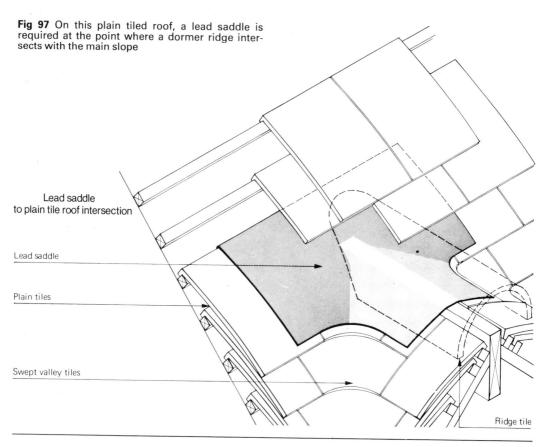

Fig 97 On this plain tiled roof, a lead saddle is required at the point where a dormer ridge intersects with the main slope

Lead saddle to plain tile roof intersection

Lead saddle

Plain tiles

Swept valley tiles

Ridge tile

When complete, parts of the saddle will be visible, but like the proverbial iceberg, a considerable part of this 'lead tile' will be hidden below the surface.

Ridge and hip rolls
A familiar sight on slate roofs is the lead roll finish on ridges and hips. As discussed in Chapter 6, concrete or clay ridge tiles are often used for weatherproofing; they are easy to install and pleasant in appearance. On the other hand, you may want to refurbish a roof to match nearby examples with lead roll ridges. To do this, you need Code 4 sheet, and must ensure that each length doesn't exceed 1.5m (4ft 11in). On the ridge these should overlap by 150mm (6in), and on the hips by 100mm (4in). Overlay on the slates should be 150mm (6in) on either side. As Fig 98 indicates, the familiar roll shape is produced by a rounded fillet which is fixed to the

Photo 110 A lead 'saddle' has to be installed at the point of intersection between the dormer ridge and the main roof prior to the addition of tiles or slates

peak of either ridge or hip. On a modern trussed roof the omission of a ridge tree means that you will need to carry out some preparatory supporting groundwork.

Wind lift may cause problems, and you should cut lead clipping strips which are long enough to fold backwards twice with a double tab over the exposed edges (*see* Fig 98). Clips must be positioned over the lapped sections, and at intervals no less than 500mm (20in). This detailing gives a familiar shape to the ridge, and it is interesting that some clay ridge tiles are purposely made to assume the traditional shape of the lead roll ridge complete with clips. If you look carefully when travelling in North Wales you may even see this roll pattern copied in large slabs of slate on hipped roofs (*see* Photo 92a, page 148).

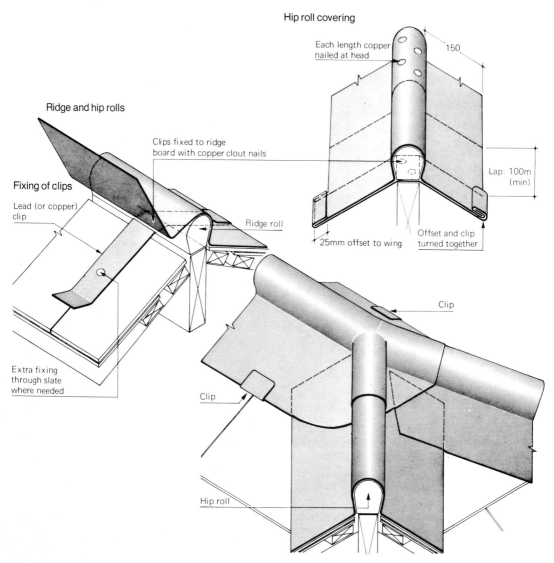

Hip roll covering

Each length copper
nailed at head

150

Lap: 100m
(min)

Ridge and hip rolls

Clips fixed to ridge
board with copper clout nails

Ridge roll

Fixing of clips

Lead (or copper)
clip

Extra fixing
through slate
where needed

25mm offset to wing

Offset and clip
turned together

Clip

Clip

Hip roll

Fig 98 Ridge and hip rolls are formed by dressing
lead sheet around a timber ridge piece shaped with
a rounded finish. Clips made from lead or copper
give added wind resistance to the exposed edges

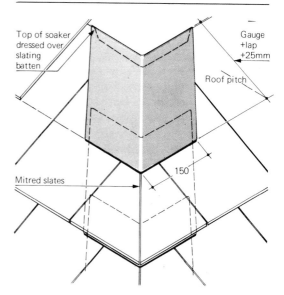

Top of soaker
dressed over
slating
batten

Gauge
+lap
+25mm

Roof pitch

Mitred slates

150

Fig 99 The mitred hip detail used on some slate
roofs is made weatherproof by the installation of
lead soakers fitted underneath the slates

184

Soakers on mitred hips

Whereas a hip covered with a lead roll or a ridge tile fulfils the objective of weatherproofing, a mitred hip is reserved for the connoisseur of slate roofing. However, some DIY builders *do* have time to attend to detail, and patience to match. No high level of skill is involved, although the ability to shape mitred slates to fit the hip slope is a prerequisite for success. Obviously the rain would penetrate the hip joint were it not for the installation of the unseen lead soakers which have to be positioned beneath the close-mitred slates. To make these, you need Code 3 lead, and it is a good idea to make the first trial pattern piece in cardboard. This will have a broad V shape not unlike the appearance of an open book (*see* Fig 99). The soaker should extend a minimum 150mm (6in) on either side of the mitred join, and the length should be the sum of gauge + lap, with a further 25mm (1in) added on for tucking over the battens meeting at the hip. Dressing the soaker over these battens means that no nailing is required. The whole operation is time-consuming, but the result is a truly striking feature (*see* Photo 92c, page 148).

Lining valleys

The advent of plastics has resulted in a decline of lead-lined valleys. There are quicker and less involved ways to produce this feature which the amateur would be advised to follow. However, if you are involved in a DIY restoration operation and insist on adopting the traditional answer, you should consult the free booklet *Lead Sheet Flashings* by the LDA. Provided the timber groundwork is sound, with solid board sole and tilting fillets on either side of the valley, this feature in lead is not difficult to complete.

Another answer is to use valley soakers – a method which has become popular on roofs covered with natural or synthetic slates. By cutting a series of overlapping soakers shaped like a kite, it is possible to form a mitred valley in slates for pitches even lower than the normal limit of 45 degrees. The result is notably attractive, and full guidance is given in *Leadwork* No 4, published January 1985 by the LDA and the BLMA.

Foil-flashing materials

The development of a self-adhesive flashing is most welcome, and there is no shortage of manufacturers currently producing foil/bitumen weatherproofing strips. However, it is important to recognise that these products are wholly different from traditional flashing materials such as lead sheeting. For example, the thickness of the metal layer is not likely to exceed 0.25mm, and its strength relies on a bituminous adhesive backing, which is around 1.25mm ($^1/_{16}$in) thick. Many of the products are made with aluminium foil, although 'Marley Flash' is unusual because it is made with real lead (*see* Photo 111).

When comparing foil flashing with lead sheet, it should be acknowledged that these strips cannot be bossed into unusual shapes. The foil can also be punctured easily, and

Photo 111 Self-adhesive bitumen-backed foils are particularly easy to cut and install. Aluminium foil is most common, although 'Marley Flash' is manufactured with a veneer of real lead

Photo 112 Flashing strips of corrugated plastic, with a hinged vertical upstand, provide a neat weatherproofing for lean-to structures – of any pitch – which have been roofed with PVC profiled sheeting

must not be regarded as a long-life performer, although it is used on some new buildings. Nevertheless, it plays an important part in repair work, and is ideal for weatherproofing lean-to roofs made with profiled sheeting. Moreover, the material has remarkable adhesion, and you will find that once pressed into place, it is rarely possible to peel it away in one piece for repositioning. It is remarkably easy to use, though you should check that surfaces are free of dust. On porous materials, a special priming paint should be applied first. When cutting the strip to shape, a pair of scissors or a sharp knife is all you require. On account of the bond which is created, problems of wind lift or moisture creep are eliminated.

If you construct a lean-to building, flashing foil can be used to form the apron cover since it does not involve intricate curvatures. Once cut to length, the foil is pressed by hand against the upstand, and on the roof covering it is easily pressed to fit around curves. As it doesn't need to be shaped by bossing, it is ideal for use with profiled sheeting such as corrugated plastic. Adhesive foil flashing is also an invaluable product for repair work (*see* page 223).

Purpose-moulded corrugated flashing

The use of adhesive flashings is one way to weatherproof plastic roofing sheets on lean-to roofs. An even neater answer is to use the tape in conjunction with the PVC purpose-made flashing pieces manufactured by Cavity Trays Ltd. As shown in the photograph (Photo 112), a short strip of corrugated plastic is moulded integrally with an upstand which has to be placed vertically against the wall. The overlay section and upstand are hinged so that the unit can suit lean-tos of any pitch. All that is needed is a cover piece on the wall, and this can be made either with flashing tape or lead sheet. Units are manufactured to suit various corrugation patterns, and on account of the standardisation of profiles set out in BS 4203 : Part 2 : 1980, it is unlikely that you will find difficulty obtaining a matching cover flashing.

Conclusion

In these chapters, the point has been made that a number of roof-covering materials are not difficult to install by the amateur self-builder. When methods of setting out and installation procedures are understood, there is every reason why a patient and thorough practitioner should embark on this operation in confidence. Certain coverings or complex roof configurations undoubtedly call for the skill of a contractor, but knowledge rather than dexterity is the order of the day on a *simple* 'lid'. After the initial marking and measuring, you will find that work takes shape at encouraging speed, and there is a splendid feeling of satisfaction as the elements are progressively excluded from your property.

8 Maintaining and Repairing Roofs

Before attempting to repair a roof, it is important to understand its construction. For instance, removing a damaged slate would be fraught with problems if you didn't realise that slate roofs are sometimes centre nailed whereas others are head nailed. In previous chapters, the constructions of many types of roof have been described, and clues to repair tasks will often be found in these earlier sections. Here the aim is to highlight repair strategies which are different from original work. The subject of re-covering roofs is also discussed; a temporary repair may prolong roof life for a while, but time comes when total replacement is inevitable.

In repairs, like construction, there are distinct differences between flat roofs and pitched roofs. Of necessity, these are treated separately.

Photo 113 Leaking glazing bars can be effectively weatherproofed with repair tapes such as Marley 'Sealtite'

Flat-roof maintenance and repair
Inspection and maintenance

Prevention is better than cure, and periodic inspection and maintenance is not usually difficult on a flat roof. In particular, three areas should receive your attention.

Drainage: check that outlets are clear and ensure that nothing obstructs a free discharge of rainwater. If ponding occurs, this may be the outcome of settlement, and you will have to increase the fall for a long-term answer. In Chapter 3, reference was made (page 56) to tapered block insulation overlay which creates a steeper roof slope without alteration to the substructure.

Surface check: clear leaves, moss and other vegetation. On a flat roof a layer of solar-reflective chippings may have been installed to reduce the heat of the sun. Check that chippings are not wearing thin, and add more if necessary. Reflective chippings can

be embedded in a fresh coating of bitumen, as long as the roof is dry.

Weatherproofing strips: upstands or strips around ventilators or roof lights may be showing signs of deterioration. Open cracks should be sealed with elastomeric sealants or proprietary bitumen compounds. Bitumen-backed flashing foil (*see* page 185) can be applied as an additional protection. If you are faced with leaking window lights, for example over a conservatory, to remove all the panes and replace putty in the glazing bars is a long job. As shown in Photo 113, sealing strips such as Marley 'Sealtite' are now available which provide a much quicker answer.

Problems

Built-up felt roofs are subject to several problems including felt damage due to wind uplift, thermal movement aggravated by solar effect, and moisture penetration induced by slow surface run-off. If a leak appears in the ceiling below, locating the source of the problem is sometimes difficult. An ingress of water may track between the layers of felt, create delamination, and make an unwelcome entry at some distance from source. If a cure is not hastily effected, problems may extend to the decking material which then starts to deteriorate. Dampness creeping between the layers of felt is hard to check, and when the top surface heats up in the sun, the trapped moisture often causes blistering.

Other signs of problems are wrinkling of the top layer caused by thermal movement, or you may notice that the entire surface looks mottled, which is a sign that the bitumen is degenerating through oxidation. Finally look out for surface puncturing, particularly if ladders are erected on the roof for window-cleaning access. Depending on severity, there are three ways to approach these problems. If the damage is localised, patching is worth a try and can produce a successful result. The accompanying illustrations (*see* Fig 100) show the simple way to deal with blisters or cracks. A heat gun, a cutting knife, a wallpaper roller, some cold bitumen compound and a patch of foil-backed bitumen flashing are the few items needed. How-

ever, these 'patch-ups' are scarcely worth while if the covering has reached an advanced state of deterioration, with faults appearing over the entire surface. In this situation you are faced either with stripping off the felt and relaying the roof, or applying a complete waterproofing cover over the entire surface.

Cover-up

The addition of a resurfacing compound as an alternative to refelting is not wholeheartedly endorsed in the trade; but it may be worth a try, and could extend the working life of the failed covering for five years or more. The antagonists claim that the addition of a sealing membrane merely holds in dampness which has accumulated between the failed layers of felt. There are vested interests on both sides of the argument, and it is noticeable that a number of chemical companies have recently launched cover sealing products. Three different cover-up systems are prominent, and the fact that several proprietary brands are marketed for DIY use is an inducement for the amateur to seek this answer to a failing built-up felt roof.

Bitumen-based compounds

A merit of this cover system is the fact that your bitumen-based built-up felt roof is being treated with compatible products. The new waterproofing coat is usually bonded with a flexible reinforcement fabric, and each coating can be applied with a paintbrush, dustpan brush or soft broom according to the size of area. Some brands, such as Aquaseal Weatherwise Heavy Duty, can be used on a damp surface – thus allowing work to be carried out in the winter. Brochures from BP Aquaseal are helpful in explaining their products, and if you have a small roof to repair, small quantity DIY repair packs are available.

A wide range of roofing repair compounds is also available under the 'FEB' brand name, and a very useful explanatory booklet entitled *Introducing the Febflex Roofing Range* is available free of charge from the FEB organisation. Many of the coatings and treatments are bitumen-based, but the company manufactures elastomeric membranes as well.

Fig 100 Procedure for repairing blisters on a felt roof:
1 Pressure build-up from water vapour beneath the felt surface can lead to blistering. This problem is aggravated if a protective layer of solar chippings is absent or in poor condition
2 After scraping away solar reflective chippings, make star cuts across each blister with a sharp cutting knife. Take care not to cut through more layers of felt than is necessary
3 The blistered area should be prised upwards along the cuts. If the felt is brittle, apply gentle heat
4 The underside of the folds, and the exposed part of the felt layers must be dried thoroughly. A heat gun is ideal for this task
5 When you are certain that there is no more entrapped moisture, coat the exposed area with cold bitumen roofing compound
6 Fold back the cut sections and make a firm bond over the repaired area with a decorators' roller
7 Complete the repair with a cover patch, held by a further coating of cold bitumen compound. Finally, add a generous layer of solar reflective chippings

Liquid rubber elastomeric membranes
Equally easy to apply are coatings which dry to form a rubberised sheet. The membrane is tear resistant and flexible, and can be applied either with a brush or spraying equipment. Several proprietary brands are available, and products like Isoflex are particularly suitable for DIY repair work. Prior to application, all loose dirt and chippings must be removed, and a primer is needed on surfaces previously treated with bitumen-based products. A thick covering is formed if several coats are applied, and, on account of the strength of the rubberised sheet, several uses other than roofing are being explored by the manufacturers.

Polyester resin membranes
Resurfacing with polyester resins reinforced by glass-fibre matting is another way to form a continuous cover. 'Glassguard' is a flat-roof repair product in this category, and if you are familiar with fibreglass car repairs or boat building you will already appreciate the

strength of reinforced polyester plastic. The special procedures applicable to roof repairs are clearly shown in the literature available from Glassguard, and, in contrast with the previous cover compounds, this material is available in several colours.

To sum up repair work, it is pleasing to report that more and more literature giving advice is being produced. Publications are usually product orientated, but their presentation acknowledges that many readers will be amateur renovators. Although technical data is included, the content is not solely addressed to architects or specifiers. One of the most comprehensive publications is *Reroofing – A Guide to Flat Roof Maintenance and Refurbishment*. This is a well-illustrated guide which can be purchased direct from Euroroof Ltd; a cover photograph shows the Roman Pantheon (built in 26BC) which has recently been re-roofed with Euroroof's high-performance cover material 'Derbigum'.

Structure and ventilation

Repairs to structure are best understood with reference to Chapter 3 and the section on decking (pages 55–6). Perhaps the chief concern is the problem of condensation and ventilation. Repairs, which may be quite serious, can be avoided if provision has been made to reduce the likelihood of condensation forming in roof voids. It is not difficult to fit ventilators in soffits, but if roof vents are needed this is a job for a roofing contractor. Check the problems identified earlier, and remember that built-up felt roofs over kitchens, bathrooms and utility rooms are particularly at risk. In the long term, your efforts now may be well rewarded in the future.

Pitching up

The point has been made earlier that low-pitched roofs have a slower discharge of rainwater, which has the effect of increasing the incidence of leaks. To avoid problems, flat roofs are not usually flat, but are built with a fall – albeit a gentle one. Yet in spite of the benefit of modern materials, the skills of modern architects and close supervision during construction, some 'flat' roofs are a constant source of problems. It is recognised in the roofing industry that in extreme cases, where repeated repairs become expensive, the ultimate long-term answer is 'pitching up'. The fact that prefabricated roof trusses can be designed to cover wide spans has produced a conversion cure for a number of buildings. Photos 28a and 28b (page 42) show properties in a northern new town which were pitched up not many years after their completion. It can be assumed that the local authority considered this cheapest in the long term, and it is a strategy open to the reader who is plagued with recurring flat-roof failure.

Pitched-roof maintenance and repair
Inspection and maintenance

Inspecting a pitched roof externally is more difficult than checking a flat roof. If there is loft access, faults can sometimes be detected from the underside, though this is unlikely if a barrier of sarking felt or board has been installed. A good pair of binoculars can sometimes help to identify weak spots externally before making a ladder inspection. In the accompanying photograph (Photo 114) of a faulty slate roof, black shadow from a gaping hole near the ridge clearly reveals a problem. In making periodic checks, attention should focus on the following key areas, which are illustrated in the following photographs (Photos 115 and 116).

Roof covering: look for cracked slates or tiles, and check if any appear dislodged. Look for an excessive growth of moss which will harbour moisture and interfere with drainage. Remedies for these problems are discussed on pages 197–98.

Hips, ridges and coping stones: on the majority of roofs, ridge tiles are not held down by mechanical fixings, and a mortar bedding offers minimal resistance to wind lift. Coping stones or hip tiles are similarly vulnerable, and pose a serious hazard if they fall from a roof. Immediate action is essential in the interests of safety.

Valleys: a valley gutter must effect a speedy discharge of rainwater and its performance is important. Bad leaks in roofs are often attributable to faulty valleys, particularly when a blockage redirects rainwater over the sides of the channel. The photo-

Photo 114 A binocular inspection from ground level reveals a gaping hole resulting from a lost slate

Photo 115 A plain tile slipping out of place is a warning that the life of the fixing nails is reaching its limit

Photo 116 Mosses and lichens may look at ease in a rural setting. But they hinder the discharge of rainwater, and lead to drainage blockages

Photo 117 Ridge tiles laid traditionally on a bed of mortar offer minimal resistance to windlift

Photo 120 A zinc flashing, corroded after nearly a century exposed to atmospheric pollutants, offers NO weatherproofing at this abutment. A replacement flashing – preferably in lead – is urgently needed

Photo 118 A dislodged coping stone must be regarded as a hazard to passers-by – not to mention the slates which first arrested its fall

Photo 119 Blockages in valley gutters lead to an unsatisfactory discharge of rainwater; the ingress of rain into the roof space is an inevitable development

graph (Photo 119) shows a blockage from detached slates and masonry, but branches and leaves cause similar obstruction.

Chimneys and abutments: another weak spot occurs at abutments, especially if the weatherproofing cover between roof and vertical surfaces is reliant on a fillet of mortar. Differential movement in the adjoining materials and settlement movements of the building cause this to fracture. But even flashing materials deteriorate, especially zinc which suffers badly in some industrial environments. As shown in Photo 120, replacements may be urgently needed. Chimney stacks are similarly a source of problems, and the condition of their brickwork needs periodic inspection. If chimney pots are held in a bedding of mortar – referred to as 'flaunching' – cracks can occur due to thermal movements.

General repairs

Several of these problems are easily rectified, and if you refer back to the previous chapters, tasks such as bedding ridge tiles, constructing hips and installing flashing have been described in the context of new roofs. Replacement work is little different, except for the need to clean away redundant mortar. Only tasks requiring different strategies are explained here.

Photo 121 An old roof covered in asbestos slates has been temporarily repaired using modern synthetic slates cut to size. But in the long term, a new covering is the only satisfactory answer

It must be recognised that some roof damage occurs through unavoidable eventualities such as a falling branch, or a freak storm. On the other hand, repeated faults portend a failing roof, and the constant need to effect repairs becomes an increasing imposition. If slates are forever slipping, this may be the sign of a 'nail-sick roof' – a term used to denote that the fixings are corroding and failing. Patching and temporary repairs effect short-term cures (*see* Photo 121), but in time, all roofs eventually need to be re-covered. In the meantime, some problems can be solved more simply.

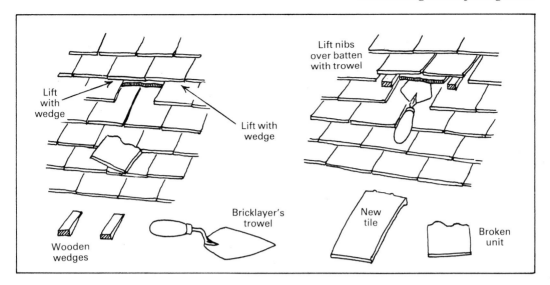

Replacement repairs

Fig 101 A trowel and a pair of wooden wedges facilitates the removal of a broken tile

Individual slates and tiles are held down by adjacent units, and a trowel is useful for prising them upwards. As shown in Fig 101, wooden wedges help to elevate neighbouring units, thus facilitating the removal of damaged remnants. If the broken piece is nailed in place, you will need a slater's rip. As shown in Fig 102, the rip is designed to locate against and force out the unwanted fixing nails which attach a slate or tile to its batten. Sometimes it is possible to sever the fixing with a long hacksaw blade, but hiring a rip is likely to be your best bet.

Nibbed tiles: as already described, modern tiles are made with nibs, but according to the degree of exposure some courses may be nailed as well. In this event, a rip is used to pull out the fixings which anchor the damaged tile. With wedges in position as shown in Fig 101 the replacement tile is slid into place. The new unit will have to rely on nibs alone since the nailing points will be covered up.

Non-nibbed tiles and slates: the impossibility of renailing a flat tile or slate in its original fashion calls for alternative fixing methods. The traditional answer was to use a narrow strip of copper or lead known as a tingle. The photograph (Photo 122a) shows how this is nailed into place prior to the insertion of the replacement unit. The protruding end of the tingle must be folded over the replacement slate to finish flush with the sur-

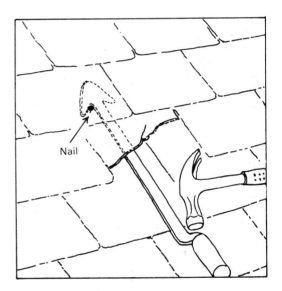

Fig 102 A slater's rip is used for removing the nails which attach slates or tiles

face (Photos 122b and 122c). This helps to resist a sliding mass of snow which will have the effect of unfolding the tingle. Though this is a long-standing method of coping with a detached unit, it is cumbersome, and the tabs which can be seen from the ground proclaim that failures have occurred. There are stories in the trade of roofers using a slate-pivoting technique made possible by single-nail fixing; slates are swung outwards from the point of damage in a fan style, and a replacement unit

197

Photo 122a Using a lead tingle to refix a slate the traditional way. A strip of lead, referred to as a 'tingle', is nailed into a batten between a perpendicular joint

Photo 122b The slate is carefully slid into position on top of the tingle, and aligned with slates in the same course

Photo 122c The lead is folded to secure the tail of the slate, and then doubled to increase its strength when the roof is covered in snow

is single nailed without the site foreman's knowledge. But single-nail fixing is *not* good practice, and a better means of attachment has not long been invented.

'Jenny Twin' slate fixer

The best inventions are often simple, and when Richard Coleman launched the 'Jenny Twin' a few years ago, the natural reaction was to wonder why no one had thought of it before. As Fig 103 shows, this is a far better way to reattach a replacement slate, and nothing is left visible on the surface. A slate has to be drilled to accept the aluminium enclosure plate, and its wire hinge is folded flat when the slate is slid into position. But as soon as it passes beyond its support batten, the hinge drops downwards to provide anchorage. The Jenny Twin is made from galvanised steel and aluminium, to assure long life, and can be used irrespective of whether the roof is built with sarking board or felt. It may take time before this product is a stock item in builders' merchants, or 'handy packed' in DIY shops. In the meantime, information on supply is available from the manufacturer direct. (*See* Appendix 2.)

Vegetation problems

There are contrasting views about the growth of moss and lichens on roofs. With permeable coverings such as hand-made clay tiles, there is a possibility that moisture from a prolific growth of moss would be absorbed and increase the risk of frost damage. This is not a problem with concrete tiles, but it remains important to ensure that vegetation does not hinder the discharge of rainwater. However, in rural environments, the growth of lichen is undoubtedly in keeping with the surroundings. Indeed, when a replacement tile or slate is installed, the new intrusion is likely to stand out as an unfortunate mismatch. To hasten a vegetation cover-up, a long-standing country tradition is to coat the surface with either skimmed milk, or a diluted wash of cow dung and water. 'High-tech' society seems to offer nothing better, and a large roofing contract in the Cotswolds was treated recently in this manner to hasten growth. It is a recognised fact that lichens

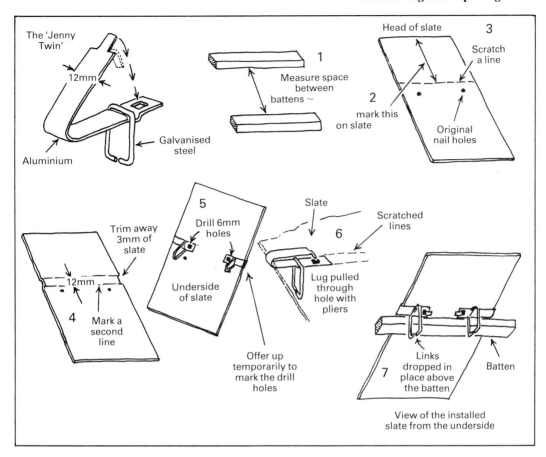

Fig 103 Procedures for re-attaching a slate using the 'Jenny Twin'

rarely grow in polluted industrial environments, but in rural settings growth generally appears on new roofs after four or five years – without the help of milk or manure.

Some growths, however, are discouraged. For example, a green deposit of algae, often seen on the northern aspect of natural- or synthetic-slate roofs, is certainly not attractive. Removing these deposits can be done by applying chemical coatings, or by fixing a copper wire to the roof. The use of toxic sprays may constitute a risk if there is nearby garden produce, and the treatment is ineffective if it rains soon after its application. You must exercise care if you use toxic chemicals, and several recipes are recommended by the major roof tile manufacturers. There are also proprietary treatments such as Febflex Fungicide (FEB [Great Britain] Ltd). This can be used on a variety of materials including metal roof coverings such as corrugated iron.

An alternative cure is to fix copper wire along high points of a roof. You may have noticed that roof tiles are always clean directly beneath a roof-level telephone wire or electricity cable. This principle is put to good effect by fixing copper wires on either side of the ridge, and rainwater passing over them contains sufficient chemical to attenuate the growth of vegetation.

Chimney repairs

Faults in chimneys can be considered in several distinct categories, and provided the access precautions detailed in Chapter 1 have been taken, repair work is not difficult.

Base of the stack: weatherproofing at the base of the stack may need attention and in severe cases replacement lead flashing is the best answer. Successful over a short term is bitumen-backed foil. Cement fillets are certainly worth replacing with flashing.

Brickwork: in its exposed position, pointing often suffers from the weather. Use a plugging chisel to chase out crumbling mortar, and dampen the joints thoroughly before repointing. A mortar mix of one part cement to five parts sand should then be pressed into the joints with a trowel. A weatherstruck pointing, in which the mortar is sloped outwards, helps to shed water (*see* Fig 104). If you have misgivings about your bricklaying skill, remember that small faults in workmanship will probably pass unnoticed at ground level.

Rendering: similar comments are applicable to rendering; if you decide to make good any damaged portions, your efforts will not receive close scrutiny. Nevertheless, the

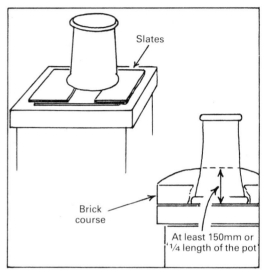

Fig 104 Brick and rendered chimney stack alternatives showing the provisions for weatherproofing which includes flaunching, oversailing, weatherstruck pointing, and the installation of damp proof course. The option of a precast concrete capping is also shown

Fig 105 Prior to the introduction of interlocking flue liners and integrated chimney units, traditional pots were supported on slates. When replacing a chimney pot, at least 150mm (6in) or a quarter length of the pot (whichever is greater), must be embedded at the top of the stack

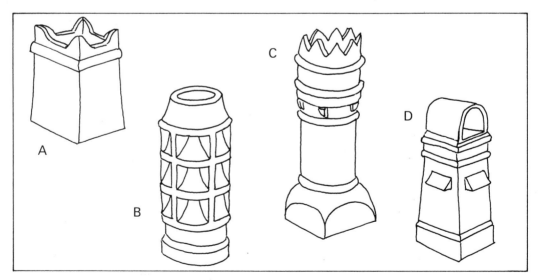

Fig 106 Examples of some of the more ornate chimney pots in the extensive range from Red Bank Manufacturing Co
A Square pot   **C** Bishop pot
B Venetian pot   **D** Hooded square with pockets

work must be sound, and you must make sure that loose material is removed, and the area dampened before applying a cement mortar mix of 1:5 of cement/sand with plasticiser. Alternatively you can use the traditional rendering of 1:1:6 cement, lime and sand. Rendering is often painted, and by recoating the entire area, patch repairs are well hidden.

Flaunching: capping mortar helps to anchor the chimney pot and shed rainwater well clear of the stack. Small cracks can be sealed with a proprietary brand of gun-applied non-setting mastic. More serious work may merit a new flaunching of 1:3 cement-sand mortar mix. Remove crumbling material carefully because the pot may be poised on a slate base; neither the pot nor mortar should be allowed to tumble down the flue (*see* Fig 105).

Chimney pots: if you decide to remove a chimney pot, don't underestimate its size. The accompanying photograph (Photo 123) shows an example of a particularly large pot which was removed as part of a DIY refurbishment scheme. Good scaffold and the help of a neighbour brought it gracefully to earth, but it was no easy task. You must be fully aware of the need for good access, good planning, and no small measure of muscle.

If you decide to renew a pot, the comprehensive catalogue from Red Bank Manufacturing is well known for its ancient and modern models (*see* Fig 106). For renovators of old properties, facsimile copies of traditional designs can be hand-fabricated to special order.

When a flue is redundant, it is a popular ploy to remove the chimney and shorten the stack. However, the flue must *not* be sealed because condensation can form inside; airbricks are needed in the stack, and it would be wise to seek comments from a building surveyor or architect before installing a capping cover.

Coating repairs

As shown in the photograph (Photo 124), leaks are sometimes checked with waterproofing surface coatings. In both Cornwall and North Wales – especially in exposed sites – a wash coat of cement is often used, but this needs periodic treatment. Frost damage and

Photo 123 Failed or redundant chimneys are often best removed. But never underestimate their size! This one took a full access scaffold, plenty of muscle, and a long weekend to bring it down to earth

thermal movement inevitably induce cracking. You will also come across 'bodge jobs', and in the same photograph you will note the bitumen and cement painted around the joints of slates in an attempt to seal weak spots. This is a dubious corrective measure which looks most unattractive.

A leaky roof can also be treated with a chemical coating strengthened by a reinforcing fabric (*see* Photo 125), and this was mentioned in connection with flat roofing. When used on pitched roofs, grave doubts are expressed about its merits, and waterproofing systems are not supported by British Standard recognition. In terms of cost, coating a failing roof is much cheaper than a re-covering operation. Against this benefit is the fact that coatings seal in residual damp in the loft

201

Photo 124 Three attempts to seal the roof below the chimney stack – a cloth coating barrier, an application of bitumen, and cement mortar pointing. None is likely to effect a satisfactory cure, and reroofing is the only long-term answer

Photo 125 Coating a roof with fabric reinforced sealants is a questionable repair strategy. Coatings are likely to crack, and condensation problems can develop within a non-breathing barrier – not to mention the unattractive appearance

space – unless new provisions are made. Moreover, problems occur if the coating gets damaged; rainwater can track underneath the layers and enter the roof at the first available weak spot. Locating the fissure can also be difficult.

On flat roofs the case for chemical coverings is much stronger, and their use is scarcely noticed. On pitched roofs, however, they are unpleasantly ugly, as the Photo 125 shows.

Re-covering pitched roofs

Many properties were built in this country around the turn of the century, and roofs were not covered in the long-lasting concrete tiles which have made their commercial début more recently. After seventy or more years, many roofs now need re-covering, and a thriving industry has developed to encourage refurbishment. There are many vested interests, and evidence of reroofing activity can be seen in any large town.

There can be no argument that repairs should always be tried first in the attempt to prolong life. But the life of a roof covering is finite, and the time comes when patching is no longer feasible (*see* Photo 126). There are a number of visible clues which herald the bad news of failure, and the accompanying photograph (Photo 127) gives a ten-point problem check. In the case of many of the older roofs finished with Welsh slate, it is often the fixing nails which have failed. A restrip is unavoidable, but on close inspection many of the slates can be reused. The chance to use reclaimed materials is particularly fortunate in areas like North Wales, where planning rules impose constraints on alternatives. However much that home owners might like to reroof with a cheap concrete tile, this break with regional tradition is disallowed – as a drive around Caernarvon soon proves.

In the case of clay tiles, it may be the coverings which have failed. In the absence of today's devices for accurate measurement of temperature, older clay tiles were sometimes baked badly. Weather damage eventually causes their surface to break up, and the photograph (Photo 128) shows a roof whose tiles are in an advanced state of delaminating or 'spalling'.

Photo 126 The life of a roof is finite. Repairs will extend its lifespan, but eventually a new 'lid' is the only answer

Photo 127 Common roofing faults – and a clear case for re-covering rather than repairing.

1. Missing slates or tiles let in rain which can cause costly damage to roof timbers apart from ruining decorations

2. Slipping slates, caused by corrosion of the fixing nails, can slide off the roof and are very dangerous

3. Valleys can be the first part of a roof to leak. Valley welts and linings can fail and movement can create cracks and holes which are difficult to repair

4. Ridges normally need attention more often than the rest of the roof. Mortar bedding can crack and the ridge tile become dislodged

5. Flashings to parapets and abutments, which are dressed into a joint in the brick work, can come loose and pull away from the wall

6. Mortar fillets are cheaper than lead flashings and are often used at abutments. But cracks are caused by movement, and frost action may lead to water penetration

7. Blocked gutters impede the flow of water from a roof

8. Copings can come loose and need to be re-bedded and repointed

9. Crumbling chimney stacks are dangerous and should be repointed

10. Cracked mortar flaunching allows chimney pots to become loose and to be dislodged

Photo 128 Flakes on the roof are evidence that the roof covering units are spalling. This problem of delamination is hastened when absorbed water freezes within the tiles, leaving a re-covering operation as the only answer

Chapter 5 evaluated different covering materials, and if a local planning department permits a degree of choice, you may want to make a change. Remember that if you select a heavier material, strengthening may be needed in the roof structure.

Sources of advice

In a competitive industry, it is no surprise that customer advisory services are notably good; most tile manufacturers publish information guides and some operate a telephone help line. Meanwhile, the slate companies also promote quarried coverings, and co-operate with each other to promote continued use of an attractive natural product. For example, if you send a small sample taken from an existing slate roof to the Penrhyn Quarry in North Wales, experts will identify its quarry of origin, and advise how to obtain a matching example. The subtle hues and variegated colours of Penrhyn are markedly dissimilar from the dark, consistent greys of Ffestiniog and the silvery-browns from Delabole quarries in Cornwall, but with co-operative effort within the home slate industry you are able to preserve an existing colour in reroofing work. Meanwhile, the facsimile 'rival' is promoted with vigour, and you shouldn't hesitate to contact companies manufacturing synthetic slates if you require advice on their products.

With the regrowth of interest in natural materials, enthusiasm for clay tiles has been revived. Quality is more consistent today with modern equipment able to monitor baking temperatures accurately. The work of the Clay Roofing Tile Council has drawn attention to the potential for clay tiles, and their publications and video presentations are re-kindling enthusiasm for this type of covering. The Council are able to advise, and steer enquiries to their member manufacturers.

The concrete-tile companies are also sensitive about service. For example, Marley Roof Tile Co publishes a free booklet, *How to Safeguard the Roof over your Head,* and operates a national network of Allied Roofing Merchants (ARM), where products can be seen on display. Redland Roof Tiles also holds a register of 'Redland Approved Re-roofing Advice Centres', and immediate advice is available by Freephone service from the Redland National Re-roofing Advice Centre. A free booklet *The Redland Guide to Re-Roofing* includes details about products and help procedures.

Lastly, the Building Centre in London and its branches throughout the country have permanent displays of products, and a free leaflet service.

Repair grants

The changing fortunes in local authority finance render it impossible to give up-to-date information about grand-aid schemes. Many initiatives have been taken at central government level in order to maintain the standard of the existing national housing stock. Reroofing is deemed to be an essential structural repair, and certain properties built before 1919 fulfil the criteria which makes them eligible for ninety per cent repair grants. Different conditions apply in Scotland, but as a general principle throughout the United Kingdom successive governments have created inducement schemes which encourage householders to embark on the refurbishment of ageing properties. Guidance on grant aid can be obtained from the manufacturers of roof tiles, but for the latest position you should consult your local authority. For a thorough review of the subject – including a section on roof strengthening procedures – you should obtain the booklet *Grant-Aided Housing Rehabilitation* published by the Redland Roof Tile Co.

Selecting a contractor

Selecting a roof repair specialist is bewildering at first sight. The potential for custom is evident if you look at the daunting listing of local contractors in the Yellow Pages. It has already been mentioned that some manufacturers carry registers of specialist contractors, and if you use one of their members this can give a guarantee period for workmanship. For example, Redland Tiles now carry a hundred-year guarantee, and if installed by a Redland-approved reroofing contractor the product is accompanied by a ten-year workmanship guarantee. It is a good idea to seek the advice of manufacturers rather than picking a contractor at random.

Another source of guidance is the National Federation of Roofing Contractors. The Federation will supply on request a regional listing of NFRC members, and their obligations will be clearly explained.

Self-help

Notwithstanding the good sense of employing an expert, there are many DIY enthusiasts who enjoy the satisfaction of self-help. If a building qualifies for a ninety per cent grant, this might be a foolish strategy since the grant covers labour as well as materials. But otherwise there are many reasons why amateur builders might choose to do the work themselves, and the majority undoubtedly do it well. It is certainly prudent to test one's ability first by reroofing a porch, bay window, or separate garage. If a small job proceeds well, it has been the aim of this book to provide the encouragement, guidance and explanation to tackle a more demanding undertaking. In many respects the work is similar to constructing a new roof, but removing the original covering is always a dirty operation. As a final reminder, check the content of Chapter 1.

Reroofing safely is dependent on the means of access, and the methods used to remove redundant materials. Acknowledge the wisdom of seeking professional advice in drawing up a specification, and the obligatory consents which must be obtained from the local authority. Lastly, the need for commitment and determination goes without question, but, assuming that you tackle any practical undertaking with tenacity, you will have already decided that only the sky is the limit.

9 Designing, Constructing and Repairing Rainwater Drainage Systems

A roof can be an effective collector of rain, and in countries where water is a precious commodity, this is purposely exploited. In Britain the situation is different, and precipitation in the form of rain or snow is regarded as something of an inconvenience. Most roofs are designed with a pitch to accelerate surface run-off, and in all modern properties this is supported by a system for collecting, transporting and dispersing the unwanted water as efficiently as possible. The purpose here is to describe design features which promote an effective system, to compare products, to explain installation work and to refer to typical repair and maintenance jobs.

Photo 129 Some tasks can be carried out from the top of a well-positioned ladder, but large replacement jobs need a proper work platform

Different forms of drainage

The focus of this chapter is on drainage systems used for domestic dwellings. No reference is made to flat-roofed industrial buildings which feature roof outlets, and rainwater downpipes situated inside the main superstructure. Similarly no reference is made to lead-lined box gutters which are typically constructed behind parapets, for example on Renaissance buildings and on the roofs of old churches and cathedrals. The focus here is on eaves gutters and downpipes, with an occasional reference to valley gutters.

With the exception of thatched buildings, all properties are equipped with a roof drainage arrangement. The thatched Almshouses at Moretonhampstead in Devon are unusual

because a large guttering has been constructed at their eaves, and other thatched buildings with guttering can be seen in villages near Exmouth. Arguably this spoils the familiar image of a thatched roof, although it is functionally desirable.

Access

The safety points stressed throughout this manual are just as important when working on roof drainage. It is true that tasks like cleaning gutter outlets *can* be performed from the top of a well-positioned ladder, and this is helped if you fit a clip-on rung platform such as the 'Standeezee' from White Seal Stairways. However, replacing a rainwater system is more of a problem. Lengths of guttering are unwieldy, and coupling-up sections at roof level is a dangerous undertaking without a tower or scaffold. It may be tempting to support runs of pipe or guttering with outstretched arms and legs, but aerial gymnastics on the top of a ladder is best left to trapeze artists.

A comparison of materials

Eaves guttering can be made in timber, and many terraced houses in Sheffield have this provision. But constant maintenance is always a problem. More recently, rainwater goods have been fabricated in asbestos cement, and this material has needed little maintenance. However, it is brittle, easily fractured and subject to the uncertainties associated with asbestos products.

For many years, lead was used for rainwater systems, and although it was attractively fashioned, cost prevented its widespread use. If you own a property with ornate lead hopper heads as shown in the photograph (Photo 130), it is interesting to note that Kestner Building Products manufacture glass-fibre replicas and will even take a copy moulding to produce identical replacements. However, on most domestic properties cast iron has been more popular on account of its strength. If you are refurbishing an older house, cast-iron systems are still available from Glynwed Foundries. Disadvantages are weight and the procedures for installation. Cutting sections to length is not easy, and

Photo 130 Renovators of older properties may be interested to know of Kestner Building Products who manufacture replica rainwater heads in glass reinforced plastic. Copies of originals can be made where a replica is required

provision must be made to reduce the incidence of corrosion. When joints are formed in the guttering, sections have to be held in place with a rust-proof bolt, and bedded in a proprietary mastic jointing compound or a mix of red lead and putty. At the time of installation, two coats of iron oxide are recommended as a base coat prior to painting, and on an exposed site you should add a bitumen coating inside the gutter. Thereafter, the paintwork must be properly maintained.

On parts of the continent, metal sheeting is often used, but whereas galvanised gutter and downpipe is commonplace in northern France, it is not popular in Britain. For the last twenty years preference in this country has been for systems made from uPVC – unplasticised Poly Vinyl Chloride. This is more commonly referred to as 'plastic', and has both advantages and disadvantages. Products are light, easily cut to size, and forming the joints in eaves gutter relies on rubber seals rather than caulking compounds. All kinds of shapes can be moulded, and fittings have become remarkably innovative. For example, Hunter Building Products recently introduced the ingenious leaf filter shown in Fig 107. Components like this would be very difficult to make using traditional materials.

Fig 107 The leaf filter downpipe unit from Hunter Building Products

A further benefit with uPVC is the fact that it is non-corrosive, and requires no maintenance. Paint is not needed for preservation, and for the sake of appearance uPVC rainwater goods are self-pigmented in a variety of colours. However, the long-term colour permanence of plastics has been less successful than originally anticipated. Brilliant whites seem to 'yellow', and shiny, black goods often degenerate into a nondescript dull grey within five or six years. It was once believed that decorative painting would never be needed, but this is becoming increasingly doubtful.

Disappointments also relate to the long-term durability of plastic; products which appear tough and flexible at the time of installation may lose their resilience and become surprisingly brittle. The photograph on page 223 (Photo 143a) shows a six-year-old downpipe which shattered like a potato crisp with

only a glancing blow. Further disappointments concern the methods of connection, and seepage at gutter joins is a common fault. Early systems took remarkable thumb strength to clip together, but leaked if the neoprene seals which were merely surface mounted were damaged or dislodged in assembly. Redesigned unions have appeared, and strap clips, which are snapped around the outside of each joint, are far easier to couple up. Neoprene seals which are held in a moulded recess are another improvement, and these latest design features are shown in Photo 140, page 217. Unfortunately this is no consolation for owners of earlier systems, and failed units may need replacing. Repairs can be made as discussed later, but the thermal movement of uPVC requires a compound of considerable flexibility. The familiar creaking sounds in hot weather are evidence of thermal movement, and for every 10°C (50°F) increase in temperature, a 10m (32ft 10in) length of gutter may gain 6mm (¼in) in length. Inevitably, this places a considerable strain on the joints in a run of uPVC eaves gutter.

Recently there has also been a growing interest in aluminium systems. Pre-coloured products are available, and one manufacturer lists seven options. Aluminium sections weigh less than a third of their equivalents in cast iron, and require minimal maintenance. Two methods of manufacture are in vogue, and guttering can be shaped on site from sheet material with a special fabricating machine. The contractor can produce eaves gutter any length up to a maximum of 244m (800ft) and the only limitation is the single choice of profile. Joints have to be made at the corners, and together with end caps these are mechanically fixed and sealed with special mastic.

Another system was introduced in 1983 by Alumasc which comprises cast aluminium gutter, die-cast fittings, and heavy-duty extruded downpipes. The choice of seven gutter profiles includes traditional patterns which are useful for refurbishment work. Alumasc rainwater goods as shown in the photograph (Photo 131) are factory finished in stove enamel, and a wide choice of colours is available. There is also the option of using gloss paint, although this involves cleaning and degreasing the surfaces, applying a primer of zinc chromate, followed by two coats of gloss paint. An undercoat is not required, and under no circumstances should lead-based primer be used.

Alumasc also manufacture an unusual system for new buildings called the Barclay Box system. The box system acts as an eaves gutter but also takes the place of a wooden fascia board. Runs are fixed directly to roofing timbers using special rafter brackets.

Aluminium systems resemble cast-iron products in appearance but require far less maintenance. The material shares the weight advantage of uPVC without the problem of thermal instability. However, aluminium has shortcomings, and corrosion is a problem in industrial areas. In order to avoid corrosion from a metal mismatch you must also ensure that all nuts, bolts, washers or screws are made of aluminium; substitution of galvanised steel is possible, but the zinc coating has short life when coming into contact with aluminium. Furthermore, aluminium cannot be used with roofs clad in copper, or in close proximity to lead. It is also important to check the content of any mastics used as sealants. Plastic compounds or bitumen-based mastics are acceptable, but problems will arise with lead- or copper-based compounds like red lead. However, manufacturers make this clear and usually supply suitable products to accompany their systems.

It is clear, therefore, that no material is perfect, and the final choice of rainwater goods is governed by a number of practical factors, not to mention personal preference.

Components in the system

Eaves gutter

Eaves gutters are manufactured in a variety of sizes and profiles which contribute to the efficiency of the system. The traditional profile used on cast-iron gutters was the 'ogee' pattern shown in Fig 108. This is still available in some modern products, especially aluminium systems.

Rather less fanciful is guttering which has a simple symmetrical curvilinear shape, and this is usually referred to as 'half round'. The addition of the word 'nominal' as a postscript confirms that it is half round in name only, and not exactly semi-circular. Generally the

Photo 131 The aluminium rainwater system from Alumask is an attractive alternative to uPVC systems

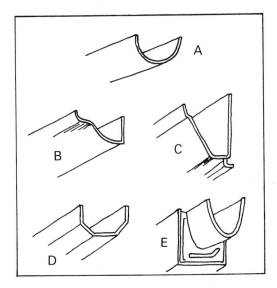

Fig 108 Eaves gutters are available in a wide range of modern and traditional profiles
A Half round
B Ogee
C Barclay Box (Alumasc)
D Squared
E Deepflow (Marley)

209

Photo 132 Marley deep profiled guttering, encased in a uPVC box, makes a neat finish by concealing fascia board completely, and copes with a greater volume of rainwater

gutter of 100mm (4in) cross-sectional measurement is less than a semi-circular profile, whereas the larger 150mm (6in) gutter is usually true half round.

Experimenting with curvilinear shapes to improve flow capacities has led to several developments. For example in 1972 Marley introduced the 'Deepflow' gutter system which could either be conventionally fixed, or concealed within a box eaves enclosure as shown in the photograph (Photo 132). With a width of 110mm (4⅜in) and a depth of 75mm (3in), the capacity of the Deepflow system is claimed to be twenty per cent greater than the 125mm (5in) true half-round gutter, and as much as sixty per cent greater when compared with the 125mm (5in) nominal half-round gutter.

Increased flow capacity is also achieved with squared profile guttering; this is complemented with square-section downpipe and is a fashionable design at present.

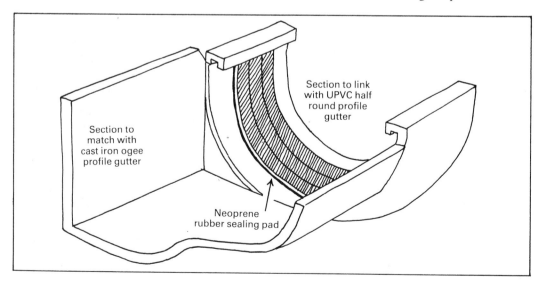

Section to link with UPVC half round profile gutter

Section to match with cast iron ogee profile gutter

Neoprene rubber sealing pad

Fig 109 Special connectors permit a uPVC rainwater system to be linked up with an existing run of cast iron guttering

Gutter adaptors

Certain circumstances dictate that new guttering is joined up with an existing eaves gutter of different pattern. For example, in terrace housing an eaves gutter forms a continuous run, and an owner deciding to replace the
section serving his property is obliged to preserve the continuity of the system. The manufacturers are aware of this, and adaptor
sections are manufactured as shown in Fig
109. Before buying a new section, check that
suitable adaptor units are available, and remember to specify whether a right- or left-
hand unit is required. Because of the problems caused by metal mismatch, you are not
advised to install aluminium guttering which
links with a run of cast-iron guttering.

Supporting eaves gutter

Methods for supporting guttering are largely
dictated by the design of the roof and its
eaves detail (*see* Fig 110). In the absence
of a significant overhang or fascia board it can
be installed with spiked rise-and-fall brackets
shown in the drawing. The flat metal spike
is driven into a mortar course, and adjustment of the supporting thread allows the
gutter to be lifted or lowered to the required
height. If a fascia board has been fitted, support brackets can be screwed directly into the

Photo 133 In the absence of a sturdy fascia board to
provide fixing points this cast iron gutter is supported by specially made brackets. These are not
out of character on this eighteenth century cottage

timber at centres (ie space intervals)
specified in the manufacturer's installation
leaflet. Rather harder to fix are brackets designed to screw directly to the rafters; side
and top fixings are available, but it is often
necessary to disturb slates or tiles in order to
expose fixing points. The Alumasc Barclay
Box system also involves rafter fixing, and
since it forms an integral part of the eaves design it is only deemed suitable for new buildings.

Valley gutter

Where two roofs meet, the intersection between their sloping surfaces is referred to as a
'valley', which like its geographical counterpart represents a watercourse. Different
methods of construction have been discussed
in earlier chapters, but the concern here is
with the discharge of rainwater into the eaves

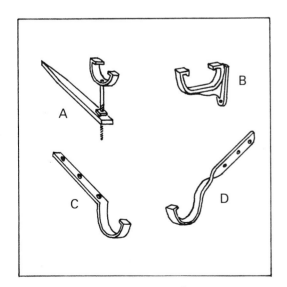

Fig 110 Supports for eaves guttering:
A Rise and fall bracket
B Fascia bracket
C Rafter bracket (top fixing)
D Rafter bracket (side fixing)

211

gutter. With short valleys serving small roof areas, it is acceptable for rainwater from the valley to discharge into an eaves gutter of adequate dimension (Photo 134). However, this should be backed up by a downpipe situated in close proximity to the junction. A long valley gutter like the example shown in Photo 135 needs better provision, and this has been made by allowing rainwater from the valley to discharge directly into a hopper head.

Outlets and downpipe

The outlets from eaves gutters are of two types. One example incorporates a 'stop end', which obviates the need for a separate stop unit but dictates exactly where the downpipe has to be situated. Right and left units have to be specified. The alternative is a 'running outlet': this accepts gutter on either side and offers greater flexibility in use.

Downpipe is manufactured to suit the dimension of the eaves gutter, and a code may be used to indicate this size match. For example, Osma refer to their systems with codes like 4½:2½ (112:68) or 6:4 (150:110). The former figure defines the cross-sectional width of the gutter in inches, millimetres in brackets, and the second is the diameter of the downpipe. However, these are 'nominal' dimensions, and the manufacturer's literature should be checked for the exact sizes.

Round- or square-section downpipe is available and an advantage of the latter is the chance to fix it tighter against a wall. If there is need to link up with an old cast-iron system, many uPVC products will fit directly into an open socket on the section to be retained. Otherwise a connector is required which will fit into the spigot (ie male end) of the existing downpipe. Most manufacturers include these adaptors in their product ranges.

Frequently a downpipe cannot follow a straight course from the eaves to its point of outflow. Altering direction requires bend connectors, and these are available in various angles. For example some of the Marley systems include 92½, 112½ and 135 degree bends which give plenty of versatility. Branch connectors are available which permit one downpipe to be fed into another as shown in Photo 136.

Photo 134 Short valley gutters may discharge into an eaves gutter of suitable size, but close proximity of a downpipe outlet to the point of entry helps avoid overloading

Photo 135 A long valley gutter must discharge into a receiving unit of adequate dimension. On this historic building, a hopper head has been used to fulfil the requirement

Photo 136 The variety of branch connectors and angle bends available in uPVC systems can cope with almost any tortuous route to a storm water drain

Swan's neck offset

An 'offset' or 'swan's neck' refers to the downpipe arrangement below overhanging eaves. In uPVC systems, this is made up on site using short lengths of cut downpipe and a pair of offset bends (*see* Fig 111). Some manufacturers recommend that the components are joined with solvent weld adhesive, whereas others advise that a push fit is sufficient. Seepage is sometimes a problem

with a push-fit dry assembly, and on large eaves where a downpipe is laid horizontally against the soffit, as shown in Fig 112, a solvent weld is essential. When downpipe has been fixed on the side wall of a building, the offset will be much longer as shown in Fig 113. If an aluminium system is installed, offsets have to be purpose-built, and in the Alumasc range, units in various sized combinations cope with a variety of eaves details and dimensions.

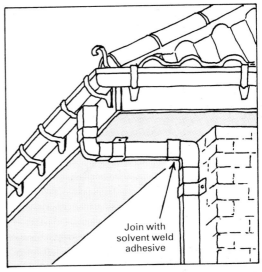

Join with solvent weld adhesive

Fig 112 A large eaves overhang may recommend the construction of a horizontal offset, but joints *must* be made waterproof with a solvent weld adhesive

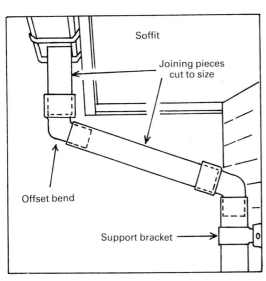

Soffit

Joining pieces cut to size

Offset bend

Support bracket

Fig 111 A swan's neck offset is made up using short joining pieces and a pair of offset bends to suit the dimensions of the eaves detailing

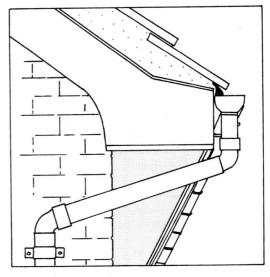

Fig 113 Downpipe mounted on a side wall may look less conspicuous, but it necessitates the construction of a long offset

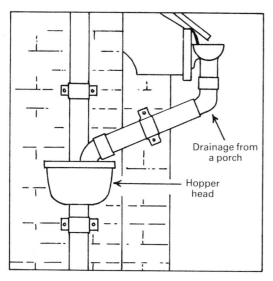

Drainage from
a porch

Hopper
head

Fig 114 The use of a hopper head permits water from other branches to discharge into the mains downpipe

Photo 137 On older properties, a hopper head was often used to act as a collecting receptacle for bathroom waste water – a practice no longer permitted

Supporting downpipes

Downpipes must be supported at intervals in accordance with manufacturers' instructions. Fixing at 1.8m–2m (5ft 11in–6ft 7in) centres is a typical specification. Different types of pipe clip are in use, and spacers are also available if a square downpipe requires an increase in wall clearance. Drive-in spikes of mild steel are produced for driving into mortar joints, but screw-in brackets are easier to fit, especially if you own a percussion drill and masonry bits.

Special downpipe fittings

On older properties a hopper head was a common feature and this allows water from several branches to discharge into the main watercourse (*see* Fig 114). When rainwater was allowed to discharge into the same drain as that used for waste water, the hopper head was often the junction point for both bathroom waste water and rainwater, as shown in the photograph (Photo 137). Nowadays this practice contravenes Building Regulations and is not used on new buildings.

Modern hopper heads are purely functional receptacles and lack the charm of the wrought-iron or lead versions often seen on historic buildings. Other special downpipe fittings are access hatches for dealing with blockages. For instance Marley rainwater goods include a downpipe unit with a screw-on sealed-cap fitting. Other makes offer removable hatches and both are most valuable if there is a risk of blockage. However, prevention is better than cure, and the time-honoured method of keeping the system clear is to fix a wire dome or 'balloon', in the gutter outlet. This component is shown in the photograph (Photo 138) and balloons can be purchased from most well-stocked builders' merchants.

Connection to underground drains

When rainwater has reached ground level, the subject of roof drainage strays beyond the limit of this manual. Suffice it to say that one method is to connect downpipe into a

Photo 139 Plastic mesh is now available, together with fixing clips, to keep gutter runs free from leaves and wind-blown debris

Photo 138 The time-honoured method for keeping gutter debris out of a downpipe is to fit a wire cage, known as a 'balloon'

100mm (4in) drainpipe via an adaptor which most manufacturers of rainwater systems produce. Alternatively, the downpipe can terminate in a fitting known as a 'shoe', which discharges the water into a gully. The former looks tidier, but the benefit of keeping the downpipe open by fitting a shoe provides easy access in the event of blockage.

Gutter grid

As shown in Photo 139, gutter grid helps to prevent leaf blockages. Most grids are made of a plastic mesh which is clipped to the top of the eaves gutters; the Coburg grid is a good example. Water discharging from the roof passes through its 5mm (3/16in) holes without difficulty, whereas leaves which settle on the top dry out and finally get blown away. It is surprising to learn that few manufacturers of rainwater goods include gutter grid in their range of products; in wooded areas this is a valuable addition to a system.

Cleaning, maintenance, repairs and replacements

Periodic clean-ups

Ensuring that drainage systems are clear of obstructions and fully operational is a necessary part of routine household maintenance. In particular, houses situated near trees will require regular attention. During hot weather, gutter sediment dries into crisp flakes, and an old paintbrush is ideal for sweeping the deposits along the gutter run. However, place a rag in the downpipe to act as a bung. Debris can be removed with a small pointing trowel, or by hand if you are not deterred by the contents. However, it is advisable to wear industrial rubber gloves; sharp edges on tiles or slates tend to graze the back of the hands which could lead to infection. When the solids have been removed, the job is completed by unplugging the downpipe and directing a hose down the run to wash away the remnants. Forcing a jet into the downpipe is another worthwhile task before returning to ground level.

Other points for attention are balloons, which should be shaken clean; if hopper heads are fitted, these have a nasty habit of becoming waste receptacles for wind-borne debris and should be cleared out.

General maintenance and repairs to eaves gutter and downpipe

Cast iron

Cast-iron gutters generally need more attention than uPVC or aluminium systems. The caulking compound for sealing joints grows brittle, and if the gutter has been disturbed the joint can fracture. Sometimes a make-shift cure is to press putty into the joint from the inside of the guttering. A more effective cure is to use a proprietary brand of sealant such as Febflex Bitumen Mastic. This is applied with a flexible knife or trowel, and the product adheres even if the surface is slightly damp. A patch-up job may seal joints with minor damage, but for bad leaks the only satisfactory answer is to saw off the gutter bolt, dismantle the faulty section and form a complete new bedding. You will need to remove all remnants of old putty, and a wire brush is useful for cleaning corroded surfaces. It is recommended to coat the exposed metal with bitumastic paint – although this is sometimes overlooked. Reforming the joint requires a non-setting mastic, such as Febflex Bitumen Mastic or Plumbers' Mait; putty can be used as a less satisfactory alternative. This is pressed into the recessed 'female' portion of the gutter, known as the 'socket', and the male portion is then positioned on the bedding material. A new galvanised nut and bolt is required to pull the sections together, and you may find it necessary to relocate or redrill the hole. Excess sealant should finally be removed, and the interior joint surfaces smoothed over to form a clear run.

With cast-iron systems a periodic maintenance task is to treat the inside of the gutter with a couple of coats of bitumen paint. Some products will adhere to damp, but not wet, surfaces; none will be satisfactory on a flaky, corroded base, and wire brushing is essential.

Procedures for dealing with cracked joints in guttering are much the same for downpipe. Joints are sometimes caulked although this is less likely if the sections interlock tightly. When making a repair, it is advisable to check the integrity of the wall brackets. The traditional fixing relied on a timber dowel driven into the brickwork to accept the fixing screws. Old plugs may need replacing with a new piece of dowel, but you might consider filling holes with new mortar and using plastic wall plugs instead.

If you find sections of guttering which are damaged, a replacement may be needed, although there are several ways to effect temporary repairs. Provided the weather is warm and dry, epoxy resins can be used, and a popular way of sheathing a cracked area is to use an old food tin. Both ends have to be removed before opening it down the middle with tin snips or a hacksaw. Having cut it to shape, you coat it with resin and place it on the outside of the faulty section. Temporary wire wound around the repair and twisted at the ends provides support while the adhesive cures. Some time later the repaired area should be liberally coated on both sides with a bituminous paint.

Cracks in cast-iron downpipe can be similarly treated, although another popular method is to use a proprietary brand of repair tape such as Sylglas. This is a fabric-based tape containing petroleum products, and the procedure is to wrap it around the damaged downpipe with an overlap at least half as wide as the tape itself. The product is very sticky, and since the pieces should be pressed into place and smoothed by hand, you must wear old clothes. Another method is to use bitumen-backed flashing foil, described on page 223.

Aluminium

Repairs to joints in aluminium guttering must only be carried out with lead-free compounds as recommended by the manufacturers. Alumasc, for example, list six proprietary makes of setting compound and four non-setting sealants in their installation leaflets. In most respects, replacement procedures are similar to those already described, except mild steel bolts must not be used to join gutter. To avoid problems from a metal mis-match, you must use bolts in cadmium, zinc or sherardised coatings. If a run of aluminium guttering is linked to a dissimilar material for unavoidable reasons, bituminous paint must be coated between the joints. During periodic maintenance, special attention should be directed to this junction point.

Plastic

Although corrosion is not a problem with uPVC systems, periodic cleaning is necessary to prevent blockages. Moreover, on some of the older systems, junctions give trouble if the rubber seals perish or become damaged. One answer is to replace faulty components, but this is not a cheap strategy. Repairs can be carried out, but thermal movements impose particular stresses on the joints. The importance of their expansion gaps is emphasised on page 229, and if these are filled with a waterproofing sealant the compound must be extremely flexible.

The accompanying sequence of photographs (Photos 142a–142j) shows preparation and repair work on a gutter run which had started to leak at the joint. After thorough cleaning, Sylglas repair tape was applied. If used on absorbant surfaces such as asbestos guttering, a primer must be used as directed by the manufacturer. This material has the flexibility necessary for temporary re-

pair work, but inevitably the expansion/contraction over a long run of eaves gutter will eventually call for replacements rather than repairs.

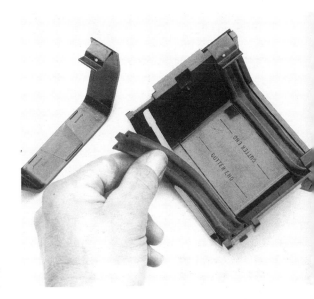

Photo 140 Design improvements on recent uPVC fittings include removeable rubber sealing strips which are held captive in a recessed slot; this is much better than the earlier face fixings which used plastic adhesive

Photo 141 The front portion of the strap clip is snapped finally into place when securing the gutter to unions or outlet components. This method of attachment is now standard on most uPVC systems

Photo 142a Gutter repair using fabric-based repair tape. Discolouration at this leaky union coupling suggests that the joint seal has failed

Photo 142b This ground floor gutter was repaired using ladder access, but hire a work platform for high-level work

Photos 142c-d Having picked a dry day, all debris was removed from the joint area with a small pointing trowel

Photo 142e The surrounding area must be thoroughly cleaned, and in dry spells, stubborn material may need shifting with a wire brush

Photo 142f All traces of dust should be removed, and an old paintbrush is ideal for 'sweeping up'

Photo 142g The repair tape, which has a cloth core impregnated with a petroleum based sealant, is cut to size on some scrap hardboard

Photo 142h Several layers of tape are offered-up to the inside of the gutter, with a generous overlap

221

Photo 142i A feather edge should be formed, using the fingers to smooth down the compound

Photo 142j The job leaves sticky fingers, and an old towel is useful for a preliminary clean-up before climbing back down the ladder

Another useful material is bitumen-backed foil which has been discussed in Chapter 7 with reference to flashing work. Its adhesive, properties are remarkable, and though not originally designed as a repair tape, it works well in this role. In the sequence of photographs (Photos 143a-143e) showing repairs to a badly damaged downpipe, the application of bitumen foil is clearly shown. Eventually a new section of downpipe will be needed, and some systems such as Osma 'Roofline' include sections with double-ended female sockets which can be introduced into the run. Meanwhile, the successful 'temporary repair' shown in the photograph is still completely waterproof five years after its completion.

Photo 143b An effective temporary repair can be made using bitumen-backed aluminium foil. This is cut to the appropriate size with a sharp cutting knife

Photo 143a Downpipe repair using bitumen-backed foil. After a few years, uPVC can become brittle; a glancing blow with a ladder caused this damage because the plastic had lost its original resilience

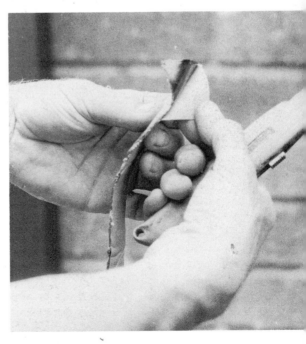

Photo 143c The bitumen layer is extremely sticky, and a backing sheet of brown paper must be removed before applying the strip

223

Photo 143d The strip must be located accurately because the bitumen backing gives a strong adhesive bond; adjustments are not easy to make

Photo 143e Once in place, pressure should be applied to the contact area surrounding the hole. This is easier in warm weather when the bitumen is pliable

Fitting eaves gutter

On account of its ease of handling, cutting and fixing, uPVC is ideal for the amateur builder, as long as there is safe access. If you are replacing an existing system there will be certain constraints, such as the siting of downpipes whose position is determined by the routeways of underground drainpipes. A first decision is whether to lay the gutter with a slight fall (ie slope), or whether to install it level. Both options are acceptable, but flow capacity and the maximum roof area that a system can drain are greater with a fall. This is worth noting, and a fall undoubtedly reduces the likelihood of overflows. In manufacturers' literature, a fall of 1:600 is recommended, and when compared to a level gutter the performance of the system is increased by at least twenty per cent. The representative fraction of 1:600 means a 1 unit drop vertically for every 600 units horizontally, for example 5mm (³/₁₆in) drop per 3m (9ft 10in) run. On a long run, however, this slope will cause the low point of the gutter to be too far below the edge of the tiles or slates. You must check that the roof drip is never greater than 50mm (2in), and on a long gutter, such as a continuous run on a line of terraced houses, the fall may need to be reduced.

Installation procedure is shown in the sequential photographs (Photos 144a-144h). Before taking fall measurements from the fascia board, your first task is to confirm if it is level. If it has been fixed with a tilt, you cannot use it as a reference for establishing falls, and will have to work with a spirit level. The photographs illustrate the installation of the outlet at the low point and the need to position the high point of the gutter as close to the slates or tiles as possible. A string line will indicate the position of gutter brackets, which in turn determine if the gutter adopts a consistent slope. When installing the system you can either decide to create a 1:600 falls by setting out levels from the high point of the run, or aim to produce as steep a fall as possible without exceeding the 50mm (2in) drip fall at the low point.

When cutting the guttering to length, make sure that it extends slightly beyond the outermost limits of the roof. This ensures

Photo 144a Procedures for fixing uPVC eaves guttering to a fascia board. With a fascia board fixing, a preliminary check confirms if the timber has been installed with horizontal accuracy

Photo 144b Taking note of downpipe position, the gutter outlet should be screwed to the fascia board, ensuring that drips from the roof will fall no further than 50mm (2in). This represents the low point in the gutter run if a fall is required

225

Photo 144c With a support bracket temporarily clipped to a section of guttering, the whole assembly is held as close as possible to the tiles at the high point of the run. The bottom of the support bracket is then marked on the fascia.

Photo 144d The support bracket is then screwed into place and a string line set up to run through to the gutter outlet. To calculate fall, measurements can be taken against the fascia board

Photo 144e The string line indicates the height required for each support bracket in order to sustain a steady fall throughout the length of the gutter run

Photo 144f It is easiest if components like 'stop ends' are fitted to the guttering at floor level

Photo 144g Before offering-up the gutter, it will be noted that on a modern property, sarking felt extends below the roof tiles. This needs to be tucked inside the guttering

Photo 144h When installed correctly, guttering complete with its stop end will extend approx 38mm (1½in) beyond the limit of the roof. This ensures that wind-blown discharge is still caught by the gutter

catchment of rainwater blown away from the corners. Careful attention must also be paid to the position of outlets; their location on the gutter run is generally influenced by the entry of the downpipe into the main drain. When assembling uPVC components, it is essential that expansion gaps are left at all junctions. As shown in the photographs (Photos 145 and 146) indicator lines are usually moulded on the socket of each fitting, and guttering must not be inserted beyond the marked limit.

Since gutter runs are unwieldy and some types of component hard to clip into place, it is often helpful to assemble as much of the system at ground level as possible. As shown in the photograph on page 227 (Photo 144f), items like stop ends can be fitted before runs are finally offered up.

Replacing downpipe

The photograph (Photo 147) shows a part-replacement job in which new guttering is matched with the original cast-iron downpipe. Adaptors are available in some product ranges, though if you are lucky a uPVC outlet sometimes fits snugly inside the downpipe as shown here.

When replacing the downpipe as well, the main drain usually dictates its position. This is especially true with a direct connection into the underground system, although most adaptor units permit some fore and aft, and side to side adjustments. More freedom is allowed if a downpipe terminates in a shoe over a gulley. However, in both cases it is wise to start the installation at the bottom; errors are easier to correct at gutter level, whereas to shift a drain involves major earth movements. All uPVC systems should be installed without any solvent welds on the downpipe run because thermal movements are accommodated at connectors. Manufacturers specify the gap needed at each spigot and socket; 6mm ($\frac{1}{4}$in) is typical.

Downpipe often has to negotiate a setback in the wall, for example where the upper part of a building has tile-hung or rendered walls. This is resolved either with a mini offset which is included in Marley's product range, or by using a pair of offset bends as shown in the photograph (Photo 148). However, at an

Photo 145 When assembling uPVC components, it is essential to ensure that the expansion gap is preserved at each junction. Manufacturers usually indicate the limit in the moulding of components

Photo 146 When this section of guttering was inserted into a coupling piece, the maximum depth line was correctly noted

Photo 147 In this partial renovation, it was fortuitous that the outlet of new uPVC guttering fitted neatly within the existing cast iron downpipe. Although a complete re-fit is the best course of action, this compromise is often possible.

Photo 148 A pair of offset bends can be used to produce a small offset – a necessary arrangement for a 'set-back' when walls are tile hung or finished in rendering

eaves overhang, a swan's neck offset must be made using two bends and an offcut of pipe. You can do this by offering up a scrap piece for the neck, with bends temporarily fitted on either end. A visual check will then show how much reduction is needed, but take care not to remove too much at a time. Alternatively the installation leaflets for some products, such as Osma and Yip, give tables to calculate the length of scrap pipe needed for different overhangs. Marley give detailed advice, and diagrams show how to negotiate overhangs of all shapes and sizes. Occasionally a deep fascia board also requires another short length of spare pipe to link the gutter outlet with the upper bend of the offset. At each junction it is essential that the spigot is on the gutter side of the flow, and if you reverse this the leaks are quite remarkable!

A support clip is needed directly below the offset, and installation instructions give further advice on the frequency of supports needed for the rest of the downpipe. If branch points are required, it seems that most manufacturers offer a variety of fittings to cope with different roof configurations.

Designing a rainwater system

On a new property, a suitable rainwater system will be specified by the architect. However, it is useful for the home owner to know something about drainage theory and to appreciate the factors which contribute to working efficiency. For example, it is not unusual for a system to overflow during periods of intense rainfall – particularly if it hasn't been cleaned for some time. If this creates a regular nuisance, the amateur builder might wonder what contributes to unsatisfactory performance and query what could be done to improve it. Furthermore, there might be an easy modification which could be made to correct a design error. Or perhaps an extension, a garage or a garden workshed has been built, and guttering and downpipe finally need to be added. Aspects of the subject are worth investigation, albeit at a relatively practical level. It is certainly helpful to be able to measure up a roof, and then know how to work out a system which would provide effective rainwater drainage with minimum expense and maximum efficiency.

Rainfall intensity

It may be surprising to learn that rainwater systems are not designed to be one hundred per cent effective, that is able to deal with every exigency of Britain's weather. Instead, systems are planned to cope with the majority of contingencies, but recognise that there will be rare overflow situations in times of freak storm. Experiments conducted by the Building Research Establishment assume a rate of rainfall of 75mm (3in) per hour. On very rare occasions an intense storm may produce as much as 150mm (6in) per hour for a short spell, but this severity is only used in calculations for buildings where an overflow situation would cause severe damage. The collected data makes no concession to region; whereas the total annual rainfall is notably higher in certain parts of the British Isles, the intensity of rainfall at any given time reveals insignificant differences.

Freak downpours happen but rarely, and information provided by the Building Research Establishment indicates the frequency which we are talking about:

> Rainfall intensity of 75mm (3in) per hour may occur for 5 minutes once in 4 years, or for 20 minutes once in 50 years.
>
> Rainfall intensity of 150mm (6in) per hour may occur for 3 minutes once in 50 years or for 4 minutes once in 100 years.
>
> Source: Building Research Establishment Digest 188, April 1976. 'Roof Drainage: Part 1', Page 1.

In recognition of these frequencies, it is usual to prescribe a system whose function will satisfy the 75mm/h (3in/h) intensity rate, and advice contained in manufacturers' leaflets is based on this figure.

Determinants of an effective rainwater drainage system

The efficiency of a system is dependent on a number of factors, such as:

The effective roof area to be drained
'Effective' roof area is not quite the same as 'actual' roof area, and this is due to the fact that wind-driven rain does not merely fall vertically, but hits a roof at an angle. The British Standards Institute in Code of Practice 308 (1974) propose that the effective roof area to receive rain equals the plan area of a roof (the area viewed vertically from above, and drawn on a plan) with the addition of half its elevation area (the area seen when viewed from the side, and given in elevation drawings). In practice, the tables drawn up by manufacturers have taken these points into account, and use the term 'actual roof area' instead of 'effective roof area'. 'Actual' roof area can be easily checked with a measuring tape. Some leaflets also make reference to roof pitches (i) below 50 degrees, and (ii) over 50 degrees, and give two sets of tables; roof pitch is linked to drive angles.

Gutter capacity
The size and cross-sectional profile of gutter contributes to the capacity of water that it can contain. This information is usually given in manufacturers' specification data, and may be an important factor when different types of gutter are compared.

Gutter fall
Water will flow much more rapidly down a sloping gutter, and it has already been mentioned that, compared with level guttering, a fall of 1:600 is at least twenty per cent more efficient in conveying rainwater towards a downpipe outlet. Some manufacturers claim rather more than this, but it is prudent to err on the side of caution.

Gutter surfaces
Friction from a corroded or rough-surfaced gutter will make a small reduction in flow rate, and increase the likelihood of holding small items of water-borne debris. Special claims are made for the Hunter uPVC gutters which feature longitudinal ribs. Increased flow rate and the reduced risk of blockages are claimed as benefits of this design.

Bends in gutter runs
A bend in an eaves gutter cuts down flow rate, although this reduction depends on how close a downpipe is situated to the bend, and whether it is sharp or round cornered.

Design of outlets
If an outlet has sharp corners, water from the

Photo 149 By placing a down pipe centrally instead of at the end of a run, a section of guttering can effectively drain a roof of twice the area

gutter will enter the downpipe less speedily. To improve the efficient removal of water, Osma produce a funnel-shaped outlet in their 'Superline' system to accelerate the discharge from gutter to downpipe.

Number of downpipes

Accepting the fact that the capacity of downpipe is designed to complement the size of its matching eaves gutter, it stands to reason that the number of downpipes will determine the speed at which water is transported away from the roof. Moreover, it also increases the area of roof which a system would be able to drain. However, cost, labour and appearance dictate that the provision of downpipes is not exceeded indiscriminately.

Position of downpipes

Placing a downpipe at the extreme end of a gutter run means that an entire 'roof-full' of water will have accumulated by the time the flow is about to pour into the outlet. However, if a downpipe is positioned in the centre of a gutter run, as shown in the photograph, the discharge will be shared equally on both sides of the outlet. This means that *no* section of gutter will have to convey more than half of the total roof discharge – although it is true that the downpipe still has to accept the same quantity of water. Effectively the roof is divided into two separate halves. Expressed another way, as long as the downpipe can cope, its central placement now means that a roof *twice* the area could be drained effectively by the guttering.

Notwithstanding the advantage of this configuration in most domestic properties, the option is not often taken because it is felt that a centrally placed downpipe spoils the appearance of a building. A preferred method for halving the flow of a gutter run is to fit two downpipes – one at each end. This is more expensive, but it allows the central part of an external wall to remain visually unbroken.

Prescribing a suitable product

When planning a system, installation literature from manufacturers is notably helpful. However, you are advised not to get too involved with flow rates which may be given, but to look at the stated roof areas which a system can serve.

When comparing products, you will note that rainwater goods are available in a variety of sizes. Manufacturers explain the intended applications; for example the Osma 6:4 Roofline product is intended for large agricultural buildings and industrial units. In contrast, the Marley 76mm (3in) Miniline gutter is ideal for a garden potting shed, but not a large house. Most manufacturers of uPVC goods present a full range to suit different structures.

When you have measured up the roof area,

you should look at the route which will be taken by the gutter (ie whether it includes bends), and where downpipes can be situated. Information on the area of roof which different products can serve should be checked in manufacturers' leaflets. The table below shows a typical presentation of data.

To illustrate procedure, the dimensions of a self-build double garage project are shown in Fig 115 and the options for a rainwater system are discussed. As usual a number of alternative designs produce an acceptable system, and making a comparative evaluation produced some interesting findings. A locally stocked product was chosen, and it was noted that the goods were available in

Fig 115 The dimensions of this roof on a self-build garage offered several alternative designs for the rainwater drainage system

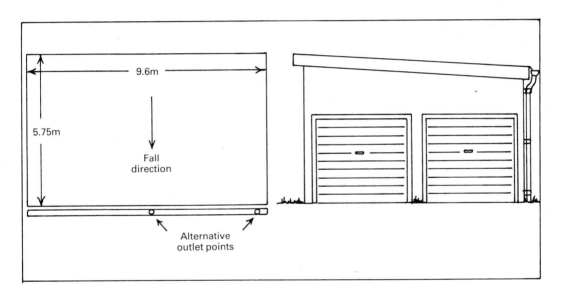

Comparative performance of rainwater goods of different sizes

Size of system	Gutter laid level	Gutter laid at 1:600
Gutter/downpipe (internal diameter) mm (in)	Max. roof area in sq m (sq ft)	Max. roof area in sq m (sq ft)
75mm/55mm (3in/2⅛in)	14 (28) sq m (150.7 [301.4] sq ft)	20 (40) sq m (215.3 [430.6] sq ft)
100mm/61mm (4in/2⅜in)	47 (94) sq m (505.9 [1011.8] sq ft)	67 (134) sq m (721.2 [1442.4] sq ft)
125mm/81mm (5in/3¼in)	69 (138) sq m (742.7 [1485.4] sq ft)	97 (194) sq m (1044.1 [2088.2] sq ft

Note: the first figure in the columns applies if the downpipe outlet is situated at the end of the gutter run.
The second figure in brackets applies if the downpipe outlet is situated in the centre of the gutter run.
Specifications vary from product to product, and these figures, though authentic, are given for example only.

three sizes. The table shows their respective performance data, together with details of the structure. A gently pitched roof of 55.2sq m (594.2sq ft) had to be served, and the underground drainage system provided two points of entry for the downpipes. The diagram shows the plan of the garage.

The table gives the following information:
75mm (3in) guttering could *not* cope with the flow – irrespective of layout

With downpipe at end of gutter

100mm (4in) guttering could *not* cope if laid level
100mm (4in) guttering would be acceptable if laid to a fall
125mm (5in) guttering could cope if laid level *or* with a fall

With downpipe in a central position

100mm (4in) guttering could cope if laid level *or* with a fall
125mm (5in) guttering could cope if laid level *or* with a fall

Several options are thus available. In view of the higher cost of the 125/81mm (5/3¼in) system, it was decided to opt for 100mm (4in) guttering and to position the downpipe centrally. The guttering could have been laid level, but a fall was preferred so that the system would be able to take extreme rainfall without any likelihood of overflow. Indeed the table data indicated that the chosen option could even cope with a roof area slightly more than twice the size.

The system was duly installed, and five years later it can be reported that there have been no failures in its performance – even in the heaviest storms.

In summary

Drainage is an important topic for any home owner, and although faults in the roof system cause less hardship than failure in soil/waste-water drainage, long-term damage to the superstructure of a property can result from leaking eaves gutter or downpipes. However, like many topics included in this manual, this is an aspect of home maintenance which an amateur can tackle with a feeling of confidence.

An innovative idea in rainwater drainage systems where a chain replaces the downpipe as this example at the Liverpool Garden Festival shows

Appendix 1: Tables

Table 1 *A comparison of weights per square metre of different types of covering materials used for pitched roofs*

Types of Covering Single lap	Manufacturer's minimum head lap	Weight in kilograms per square metre	Example of product producing this data
Trough tile (concrete)	75mm (3in)	45	Marley 'Ludlow Major'
Flat interlock tile (concrete)	75mm	50.6	Redland 'Stonewold'
Pantile (concrete)	75mm	47	Marley 'Mendip'
Double Roman (concrete)	75mm	43.9	Redland 'Double Roman'
Double Roman (clay)	75mm	41	Sandtoft 'Double Roman'
Pantile (clay)	60mm (2⅜in)	44	Sandtoft 'Pantile'
Double lap plain tiles			
Plain machine made clay	65mm (2½in)	67	Rosemary 'Plain clay tiles'
Plain hand made clay	65mm	76.2	Keymer 'Plain clay tiles'
Plain concrete	65mm	76	Marley 'Plain tile'
Slates			
Imitation Cotswold Slates (concrete)	80mm (3⅛in)	82	Bradstone 'Cotswold'
Imitation Yorkshire Slates (concrete)	80mm	104	Bradstone 'Moordale'
Imitation Slates in plastic/cement (size 610 × 305mm)			
(Methacrylate polymer/concrete)	75mm (3in)	25.78	Anglia 'Synthetic Slates'
Cement fibre imitation slates (size 600 × 300mm)	70mm (2¾in)	18.8	TAC Thrutone Slates
Quarried Welsh Slate (size 610 × 305mm)	75mm	27.9	Penrhyn natural slate

Note   (i)   This information is compiled from product data sheets, but manufacturers usually state that the figures quoted should be regarded as approximate weights per kg/m². They will be subject to small variations.

        (ii)   Where head lap is increased, the weight per square metre will also increase. Some specification leaflets give data for different head laps.

Table 2 *Recommended batten sizes for pitched roofs*

Roof covering materials	Recommended sizes			
	450mm (18in) span		600mm (24in) span	
	Width	Depth	Width	Depth
Slates:	mm / in	mm / in	mm / in	mm / in
Sized	38 1½	19 ¾	38 1½	25 1
Random	50 2	25 1	50 2	25 1
Cement fibre	38 1½	19 ¾	38 1½	25 1
Clay and concrete tiles				
Plain, double lap	32* 1¼	19 ¾	32* 1¼	25 1
Single lap	38 1½	22 ⅞	38 1½	25 1

Table based on information contained in BS 5534 : Part 1 : 1978, AMD 3554 May 1982 and manufacturers' product literature.
For further details, consult this British Standards Institution, Code of practice for Slating and Tiling manufacturers' recommendations.

*This dimension in particular varies according to product, eg Dreadnought clay tiles are given a minimum size of 25mm (1in) × 19mm (¾in) for 450mm (18in) rafter spacings; Marley concrete tiles are given a minimum recommendation of 38mm (1½in) × 19mm (¾in) for 450mm (18in) rafter spacings and 38mm (1½in) × 25mm (1in) at 600mm (24in) rafter spacings.

Table 3

Minimum recommended head laps for Welsh slate (based on BS 5534 Tables 4 and 5)

Slate size (nominal)		Moderate exposure: driving rain index less than 7m²/s — Minimum rafter pitch									Severe exposure: driving rain index 7m²/s or more — Minimum rafter pitch								
mm	inches	20°	22½°	25°	27½°	30°	35°	40°	45°	85°	20°	22½°	25°	27½°	30°	35°	40°	45°	85°
660 x 355	26 x 14	130*	105	90	80	75	75	65	65	—	—	140	120	115	105	85	75	65	—
610 x 355	24 x 14	115	105	90	80	75	75	65	65	—	145*	125	105	95	90	75	75	65	—
610 x 305	24 x 12	130*	115	90	80	75	75	65	65	—	—	120	120	115	110	90	80	70	—
560 x 305	22 x 12	115	105	90	80	75	75	65	65	—	140*	120	105	100	90	75	75	65	—
560 x 280	22 x 11	120*	110	90	80	75	75	65	65	—	145*	130	110	105	100	85	75	65	—
510 x 305	20 x 12	115	105	90	80	75	75	65	65	—	—	—	100	95	85	75	75	65	—
510 x 255	20 x 10	125*	110	90	80	75	75	65	65	50	—	—	115	110	100	90	75	65	65
460 x 305	18 x 12	115*	105	90	80	75	75	65	65	50	—	—	110	95	85	75	75	65	65
460 x 255	18 x 10	125*	110	90	80	75	75	65	65	50	—	—	115*	110	100	85	75	65	65
460 x 230	18 x 9	125*	115*	100	80	75	75	65	65	50	—	—	120*	115*	105	95	85	65	65
405 x 305	16 x 12	—	—	—	80	75	75	65	65	50	—	—	—	—	90	80	75	65	65
405 x 255	16 x 10	—	—	—	75	75	75	65	65	50	—	—	—	—	100	95	75	65	65
405 x 230	16 x 9	—	—	—	85	75	75	65	65	50	—	—	—	—	100	100	85	65	65
405 x 205	16 x 8	—	—	—	90	75	75	65	65	50	—	—	—	—	105	100	90	65	65
355 x 305	14 x 12	—	—	—	80	75	75	65	65	50	—	—	—	—	75	75	75	65	65
355 x 255	14 x 10	—	—	—	80	75	75	65	65	50	—	—	—	—	80	75	75	65	65
355 x 230	14 x 9	—	—	—	80	75	75	65	65	50	—	—	—	—	85	80	75	65	65
355 x 205	14 x 8	—	—	—	80	75	75	65	65	50	—	—	—	—	90	85	75	65	65
355 x 180	14 x 7	—	—	—	80	75	75	65	65	50	—	—	—	—	95	90	80	65	65
305 x 255	12 x 10	—	—	—	80	75	75	65	65	50	—	—	—	—	75	75	75	65	65
305 x 205	12 x 8	—	—	—	80	75	75	65	65	50	—	—	—	—	75	75	75	65	65
305 x 150	12 x 6	—	—	—	80	75	75	65	65	50	—	—	—	—	85	80	75	65	65
255 x 255	10 x 10	—	—	—	80	75	75	65	65	50	—	—	—	—	75	75	75	65	65
255 x 205	10 x 8	—	—	—	80	75	75	65	65	50	—	—	—	—	75	75	75	65	65
255 x 150	10 x 6	—	—	—	80	75	75	65	65	50	—	—	—	—	75	75	75	65	65

Note: the actual pitch at which the slate lies on the roof is less than the rafter pitch by an amount which is a function of the slate thickness and the lap. Therefore pitches marked* are not suitable for use with extra heavy slates.

Reproduced from Penrhyn and Buttermere Slate Roofing Manual with permission.

Appendix 2
Address List

Although this manual includes theoretical background information, its principal objective is to give guidance to the practitioner. Recognising this intention, both the text and illustrations make reference to specific products as well as sources from which you can obtain further information. To enable the reader to follow up the references, this address list is provided as an integral part of the manual.

Whereas some products receive repeated mentions, this list presents them in the order in which they first appear in the book.

National Association of Scaffolding Contractors,
National Federation of Building Trade Employers,
82 New Cavendish Street,
London,
W1M 8AD

Publishers of *Scaffolders and Users Guide to Safe Access Scaffolding*

B. C. Barton & Son Ltd,
PO Box 67,
Granville Iron Works,
Hainge Road,
Tividale,
Warley, West Midlands,
B69 2NJ

Manufacturers of the **Multiscaf** access tower

Burton Wire & Tube Co Ltd,
Unit 2,
Midland Road Industrial Estate,
Midland Road,
Swadlincote,
Burton-on-Trent,
DE11 0AN

Manufacturers of the **Ladder Stand Off** and the **Universal Roof Hook**

White Seal Stairways Ltd,
88 Hopewell Drive,
Chatham,
Kent,
ME5 7NL

Manufacturers of the **Standeezee** ladder platform

Sumaco Merchandising Ltd,
Suma House,
Huddersfield Road,
Elland,
West Yorkshire,
HX5 9AA

Importers and distributors of the Hirsh **Workgrabber** and **Iron Horse**

Ready Scaffolding Ltd,
5A Lombard Road,
Merton,
SW19 3TZ

Manufacturers of the **Chimneydeck** chimney scaffolding

Europalite Ltd,
Europa House,
Sheepbridge Lane,
Mansfield,
Nottingham,
NG18 5DX

Manufacturers of the **Europachute** site waste disposal chute

SGB Group PLC,
23 Willow Lane,
Mitcham,
Surrey,
CR4 4TQ

Manufacturers of the **Snaplok** roof rig

Appendix 2 Address List

Mustang Tools Ltd,
Aylsham,
Norwich,
Norfolk,
NR11 6EL

Suppliers of **specialist roofing tools** and **tile cutters**

Dimos Marketing (UK) Ltd,
Sunningdale Walk,
Greenhill,
Herne Bay,
Kent

Suppliers of **specialist roofing tools, tile cutters** and **lead dressers**

Brian Hyde Ltd,
Stirling Road,
Shirley,
Solihull,
West Midlands,
B90 4LZ

Suppliers of **specialist roofing tools** and **access equipment**

Rabone Chesterman Ltd,
Whitmore Street,
Birmingham,
B18 5BD

Manufacturer of **clinometer rules** for measuring roof pitch

H.M.S.O. Publications Centre,
51 Nine Elms Lane,
London,
SW8 5DR

Publishers of a booklet prepared by the D o E, the Welsh Office, and the Central Office of Information, 1981 entitled *Planning Permission – A Guide for Householders*

British Standards Institution,
2 Park Street,
London,
W1A 2BS

For further information on the publications giving **British Standard codes of practice**

PVC Sheet Sales Office,
Weston Hyde Products Ltd,
PO Box 15,
Hyde,
Cheshire,
SK14 4EJ

Manufacturers of **Novolux** rigid PVC roofing sheets and **Heavy Duty Novolux** made from uPVC. Publishers of DIY brochure and supporting plans entitled *Novolux for Improvement to your Home and Garden*

Timber Research & Development Association,
Stocking Lane,
Hughenden Valley,
High Wycombe,
Buckinghamshire,
HP14 4ND

Specialist association publishing *Standard Design Sheets* for roof designs, Wood Information Sheet *Principles of Pitched Roof Construction*, and literature on building timber

BIP Chemicals Ltd,
Filon Products,
Aldridge Road,
Streetly,
Sutton Coldfield,
West Midlands,
B74 2DZ

Manufacturers of **Filon** transluscent GRP profiled sheeting

OFIC (GB) Ltd,
Eardley House,
4 Uxbridge Street,
Farm Place,
Kensington,
London,
W8 7SY

Manufacturers of **Onduline** bitumen fibre corrugated roofing sheets and **Bardoline** self-adhesive bituminous shingles

Metra Non-Ferrous Metals Ltd,
Pindar Road,
Hoddesdon,
Hertfordshire,
EN11 0DE

Manufacturers of **Metiflash** zinc/lead alloy flashing material

Cavity Trays Ltd,
Administration Centre,
Pen Mill Trading Estate,
Yeovil,
Somerset,
BA21 5HU

Manufacturers of **Type X preformed cavity trays** incorporating attached lead flashing for gable abutments; **Type F corrugated flashing unit** for plastic roof sheets; **lead slates** with vent outlets. Fabricators of leadburned flashings to special order

D. Anderson & Son Ltd,
Stretford,
Manchester,
M32 0YL

Publishers of the free booklets *Built-up Roofing in the UK* and *High Technology Waterproofing for Roofs*. Manufacturers of DIY **Hippo** range of roofing materials, cold adhesive, high performance and traditional built-up roofing systems, torch-applied membranes, Solabar paint and metal-faced membranes. Complementary paints, mastics, and roof ventilators are also manufactured

Permanite Ltd,
Mead Lane,
Hertford,
Hertfordshire,
SG13 7AU

Publishers of the free booklet *Fixing Bitumen Roofings using Aquatex Roofing Felt Adhesive*. Manufacturers of bitumen roofings and underlays, **Britorch** and **Torflex** torching roofing systems, slaters felts, **Superflex** and **Hyparoof** high specification roofing system, **Permaband** self-adhesive flashing strip, **Aquatex** cold compound felt adhesive, and all ancillary products for flat roofs

Ruberoid Building Products Ltd,
Brimsdown,
Enfield,
Middlesex,
EN3 7PP

Publishers of the free booklet *Ruberoid Built-up Roofing for the Builder*. Manufacturers of bitumen roofings and underlays, **Ruberglas** 105 slates, **Rubertherm** polystyrene block and felt insulant, torching membranes, slaters' felts, high performance roofing felts, aluminium **Ruberflash,** fixing compounds, solar reflective paints and all ancillary products for flat roofs

Euroroof Ltd,
Denton Drive,
Northwich,
Cheshire,
CW9 7LU

Publishers of *Re-roofing – A Guide to Flat Roof Maintenance and Refurbishment*. Manufacturers of flat roof materials including **Derbigum** high performance roofing, **Torchtite** roofings, **Korkplus** urethane roof insulation, **Korktaper** fall system, **Bishore** paving supports, **Roofpave** tiles, **Toptrim** roof edge trim in GRP, **Icovent** exhaust vents, and all ancillary products for flat roofs

Coolag Purlboard,
Heysham Works,
Middleton,
Morecambe,
Lancashire,
LA3 3PP

Manufacturers of composite built-up roofing boards with integrated insulation such as **Coolag** standard roofboard, **Corkbond** fully bonded composite board featuring cork, **Purldek** combined plywood/insulation/vapour barrier for warm roof construction, together with allied flat roof covering materials and insulants

Catnic Metal Products,
Catnic Components Ltd,
Pontgwindy Estate,
Caerphilly,
Mid Glamorgan,
CF8 2WJ

Manufacturers of galvanised steel **joist han-**

gers, wall plate **anchor straps,** anchor straps for pitched roofs and gable ends, **roof truss clips, herringbone joist struts, Render Stop beading,** and roof space **ventilators**

Spandoboard Insulations Ltd,
Bankhall Buildings,
Stanley Road,
Liverpool,
L20 2HH

Manufacturers of **Spandoboard,** tapered insulation board designed to create falls on flat roofs

Aluminium Developments,
Bay 1,
Westgate,
Aldridge,
West Midlands,
WS9 8DJ

Manufacturers of **Alutrim** extruded alloy roof edging for asphalt and felt covered roofs, **Alugutter,** extruded aluminium rainwater system, and polypropylene 'T' breather vents for vapour pressure relief on built-up felt roofs

Gang-Nail Ltd,
The Trading Estate,
Farnham,
Surrey,
GU9 9PQ

Publishers of the free booklet, *Trussed Rafter Construction & Specification Guide.* Originators of **Concept 2000,** a micro-computer based system for the design, estimation, and manufacture of roof trusses. Manufacturers of **Gang-Nail** connector plates and specialist supports for roof trusses, backed by the Gang-Nail Timber Consultancy service

Hydro-Air International (UK) Ltd,
Midland House,
New Road,
Halesowen,
West Midlands,
B63 3HY

Manufacturers of the **Hydro-Air trussed rafter system,** supported by full consultancy and advisory services. Manufacturers of **Hydro-Nail Posi-Tooth** punched metal plate timber fasteners

International Truss Plate Association,
PO Box 44,
Halesowen,
West Midlands

Publishers of the free booklet *ITPA Technical Handbook*

The Trussed Rafter and Fabricators' Association,
24 Bishops Walk,
Gunton St Peter,
Lowestoft,
Suffolk,
NR32 4JN

Specialist trade association publishing a free leaflet describing the trussed rafter industry

Cowley Structural Timberwork Ltd,
The Quarry,
Grantham Road,
Waddington,
Lincoln

Specialists in **trussed rafters,** purpose designed **laminated roof beams,** and unusual **timber roof structures**

Kingston Craftsmen (1981) Ltd,
Londesborough Street,
Hull,
North Humberside,
HU3 1DS

Structural timber engineers with **specialist roof systems** for prestige dwellings. Comprehensive technical, design, and estimating services

The Marley Roof Tile Co Ltd,
London Road,
Riverhead,
Sevenoaks,
Kent,
TN13 3BR

Manufacturer, supplier, and installer of **concrete roof tiles.** Full support offered by a technical advisory service. Manufacturers of the **Dry Fix** roof system incorporating eaves and ridge ventilator units. Publisher of the free booklet *How to safeguard the roof over your head*

Marley Extrusions Ltd,
Lenham,
Maidstone,
Kent,
ME17 2DE

Manufacturers of rainwater goods in various patterns, including the **Deepflow** box eaves gutter, and the **Miniline** system for sheds and small DIY projects. Manufacturers of **Marley uPVC fascia, soffit, bargeboard,** and **roof ventilation systems**

Redland Tiles Ltd,
Head Office,
Reigate,
Surrey,
RH2 0SJ

Manufacturers of **concrete roof tiles,** the **Dry Ridge** system, **RedVent** roof space ventilation system, cloaked verge system, and associated roofing products. Publishers of several free booklets including *The Redland Guide to Re-Roofing.* Customer support services include freephone link with the Redland Roofing Advice Centre at Head Office

Anchor Roof Tiles (Leighton Buzzard) Ltd,
Broomhills Road,
Leighton Buzzard,
Bedfordshire,
LU7 8ER

Manufacturers of **concrete tiles** and the **Anchor** Dry Verge Fixing System

E. H. Bradley Building Products Ltd,
Okus,
Swindon,
Wiltshire,
SN1 4JJ

Manufacturers of **Bradstone** reconstructed stone slates and publishers of an instructional leaflet showing the setting out procedure for roofing with diminishing courses

Robert Abraham Ltd,
43 Bankhall Street,
Liverpool,
L20 8JF

Manufacturers of **Hardrow** concrete slates made to resemble natural stone roofing materials

Hawkins Tiles (Cannock) Ltd,
Longhouse Works,
Watling Street,
Cannock,
Staffordshire,
WS11 3BJ

Manufacturers of **Hawkins** Sandfaced tiles

The Architectural Press Ltd,
9 Queen Anne's Gate,
London,
SW1H 9BY

Publishers of **Specification '85,** from which John Duell's comments on fire hazard and thatch are cited

Thatching Advisory Services Ltd,
29 Nine Mile Ride,
Finchampstead,
Nr Wokingham,
Berkshire,
RG11 4QD

Consumers' Association,
14 Buckingham Street,
London,
WC2N 6DS

Members of the Consumers' Association can write for a register listing the Secretaries of local Master Thatchers' Associations

The Information Officer,
Council for Small Industries in Rural Areas (CoSIRA),
Queens House,
Fish Row,
Salisbury,
Wiltshire

CoSIRA hold records and addresses of thatchers in different localities

The Clay Roofing Tile Council,
Federation House,
Station Road,
Stoke-on-Trent, ST4 2TJ

For details about clay tiles and the addresses of their member manufacturers

Swallows Tiles (Cranleigh) Ltd,
Bookhurst Hill,
Ewhurst Road,
Cranleigh,
Surrey,
GU6 7DP

Manufacturers of **Swallows** hand made clay tiles

Keymer,
Nye Road,
Burgess Hill,
Sussex,
RH15 0LZ

Manufacturers of **Keymer** hand made clay tiles

Rosemary Brick & Tile Co Ltd,
Haunchwood-Lewis Works,
Cannock,
Staffordshire,
WS11 3LS

Manufacturers of **Rosemary Roofing Tiles,** smooth faced, machine made plain clay roofing tiles

Hinton, Perry & Davenhill Ltd,
Dreadnought Works,
Pensnett,
Brierly Hill,
Staffordshire

Manufacturers of **Dreadnought** plain clay smooth faced, sand faced and hand made roofing tiles

London Chemical Co Ltd,
Batchworth Island,
Church Street,
Rickmansworth,
Hertfordshire,
WD3 1JQ

Sole distributors of **Anglia** slates manufactured in methacrylate polymer concrete

Sandtoft Tileries Ltd,
Sandtoft,
Doncaster,
South Yorkshire,
DN8 4SY

Manufacturers of concrete and **Goxhill** clay roofing tiles

Decra Roof Systems (UK) Ltd,
Crompton Way,
Manor Royal,
Crawley,
Sussex,
RH10 2QR

Distributors of **Decra** lightweight steel roofing tiles

Planox Ltd,
175a High Street,
Beckenham,
Kent,
BR3 1AH

Distributors of **Isola** glassfibre reinforced shingles, **Filmtex** multidensity polythene tile roof underlay, and system **Platon** ventilating damp proof membrane underlay for turf roofs

Red Bank Manufacturing Co Ltd,
Measham,
Burton-on-Trent,
Staffordshire,
DE12 7EL

Manufacturers of **ancient and modern style chimney pots, flue linings,** and **decorative ridge tiles.** Red Bank also operate a bespoke service for the production of customer designed chimney pots

John Caddick & Son Ltd,
Spoutfield Tileries,
Stoke-on-Trent,
Staffordshire,
ST4 7BX

Manufacturers of **decorative and conventional ridge tiles and chimney pots**

Stanley Brothers Ltd,
Croft Road,
Nuneaton,
Warwickshire,
CV10 7ED

Manufacturers of **ridge tiles** and **chimney pots**

Delabole Slate Ltd,
Pengelly House,
Delabole,
Cornwall,
PL33 9AZ

Showroom, sales and information centre for **Cornish roofing slate** from the Delabole quarries

Penrhyn Quarries Ltd,
Bethesda,
Bangor,
Gwynedd,
LL57 4YG

Welsh natural roofing slate in a wide range of sizes from the Bethesda quarries; helpline and slate recognition service in operation from the Head Office

J. W. Greaves & Sons Ltd,
Llechwedd Slate Mines,
Blaenau Ffestiniog,
Gwynedd,
LL41 3NB

Welsh natural roofing slate

North Wales Slate Quarries Association,
Bryn Llanllechid,
Bangor,
Gwynedd,
North Wales

Specialist trade association concerned with Welsh slate

Burlington Slate Ltd,
Cavendish House,
Coniston,
Cumbria,
LA21 8ET

Cumbrian (Lakeland) natural roofing slate

Eternit TAC Ltd,
Meldreth,
Nr Royston,
Hertfordshire,
SG8 5RL

Manufacturers of **rainwater systems, profiled sheeting,** and **roofing slates.** The fibre cement slates include the **Eternit 2000** range, and **TAC Duracem Thrutone** range formerly manufactured separately by Eternit Building Products Ltd, and TAC Construction Materials Ltd

Tunnel Building Products Ltd,
Tunnel Estate,
West Thurrock,
Grays,
Essex,
RM16 1EJ

Manufacturers of **Tunnel II** cement fibre slates

Nicholl & Wood Ltd,
Netherton Works,
Holmfield,
Halifax,
West Yorkshire,
HX3 6ST

Manufacturers of **Zamba** aluminium slate roof and tile roof ventilators

Willan Building Services Ltd,
2 Brooklands Road,
Sale,
Cheshire,
M33 3SS

Manufacturers of **Glidevale** roof space ventilators, sealed loft access trap, and dry verge system for slate roofs

Timloc Building Products Ltd,
Rawcliffe Road,
Goole,
North Humberside,
DN14 6UQ

Manufacturers of eaves and soffit ventilators including the **Timloc Push In** units, and **Timloc loft access doors**

Westbrick Plastics Ltd,
Edison Road,
Churchfields,
Salisbury,
Wiltshire,
SP2 7PA

Manufacturers of roof ventilators including **Westbrick** eaves units designed to hold soffit boards

Energy Efficiency Office,
Room 1312,
Thames House South,
Millbank,
London,
SW1P 4QJ

For the booklet **Make the Most of your Heating** which gives step-by-step details on the installation of loft insulation

Dale Hardware Ltd,
Knowl Warehouse,
Dale Street,
Ossett,
West Yorkshire,
WF5 9HJ

Soffitex continuous eaves ventilator manufactured in galvanised steel

Black & Decker Ltd,
Westpoint,
The Grove,
Slough,
Berkshire,
SL1 1QQ

Manufacturers of the **DN538SE scroll-type jigsaw** which will form cut-outs in the soffit to accept ventilator units

Skil (Great Britain),
Fairacres Industrial Estate,
Dedworth Road,
Windsor,
Berkshire,
SL4 4LU

Manufacturers of **Auto Scroller** jigsaws which will form cut-outs in the soffit to accept ventilator units

Peugeot Power Tools,
AEG-Telefunken (UK) Ltd,
217 Bath Road,
Slough,
Berkshire,
SL1 4AW

Manufacturers of the **Peugeot 10TSE electric jigsaw.** This has a unique facility for mounting blades sideways – a useful feature when making cut-outs in restricted places such as soffit boards

Rytons Ventilation Equipment Ltd,
68 Rockingham Road,
Kettering,
Northamptonshire,
NN16 8JU

Manufacturers of **Ryton's prefabricated soffit boards** with integral ventilation units

Gyproc Glass Fibre Insulation Ltd,
Whithouse Industrial Estate,
Runcorn,
Cheshire,
WA7 3DP

Manufacturers of **Gypglas** insulating material with foil faced backing

Lead Development Association,
34 Berkley Square,
London,
W1X 6AJ

Joint publisher with the British Lead Manufacturers' Association of the free booklet, *Lead Sheet Flashings.* Publishers of *Lead Sheet in Building* by Charles Knight and Richard Murdoch which can be purchased direct from the Association

Marley Waterproofing Products Ltd,
PO Box 17,
Otford,
Sevenoaks,
Kent,
TN14 5EW

Manufacturers of **Marley Flash** self adhesive lead foil flashing material

J & D Raynes & Sons Ltd,
Lissadel Street,
Frederick Road,
Salford,
Manchester,
M6 6BR

Suppliers to the roofing industry and manu-facturers of **Rayflash** bitumen backed aluminium flashing foil. Distributors of the **Compact Angle Finder clinometer** with graduated dial

Evode Roofing Ltd,
Common Road,
Stafford,
Staffordshire,
ST16 3EH

Manufacturers of **Flashband** aluminium faced self adhesive sealing strip for instant waterproof repairs

Expanded Metal Co Ltd,
PO Box 14,
Longhill Industrial Estate (North),
Hartlepool,
Cleveland,
TS25 1PR

Manufacturers of **Expamet external Render Stop Type II** which can provide anchorage for lead flashing at the lower edge of a rendered abutment

BP Aquaseal Ltd,
Kingsnorth,
Hoo,
Rochester,
Kent,
ME3 9ND

Manufacturers of **wood preservative** and **timber protection chemical treatments** for dry rot, wet rot, and woodworm in roof and other timbers. Manufacturers of **Aquaseal adhesive flashing, roofing felt adhesives** and protective **waterproof coatings** for flat and slate covered pitched roofs. Publishers of free booklet *Aquaseal – for damp-proofing & water-proofing treatments*

Unibond-Copydex Ltd,
Yorkdown Industrial Estate,
Stanhope Road,
Camberley,
Surrey,
GU15 3DD

Manufacturers of repair products for flat roofs including **Stop Leak** instant repair kit, **Unibond** aluminised coatings, and **Liquid Rubber** pliable sealant

FEB (Great Britain) Ltd,
Albany House,
Swinton Hall Road,
Swinton,
Manchester,
M27 1DT

Manufacturers of a wide range of roofing re-pair compounds including **Febflex Bitumen Mastic** for gutters, skylights, and corrugated sheeting, **Febflex Fungicide** for killing and discouraging moss, fungal growth etc and **Febflex Hyguard** vapour permeable liquid plastic membrane. Full product review pub-lished in the free booklet *Introducing the Febflex Roofing Range*

Isoflex Ltd,
Avon House,
12 Brassmill Lane Trading Estate,
Bath,
BA1 3JF

Manufacturers of **Isoflex** liquid rubber roofing sealant for DIY and professional re-pairs on all types of roofs

Cementone-Beaver Ltd,
Tingewick Road,
Buckingham,
MK18 1AN

Manufacturers of **bitumen mastic, gritting compound** for bonding aggregates to felt roofs, **gutter sealants,** and a variety of repair products listed in Cementone's publication, *Caring for Roofs*

Wailes Dove Bitumastic plc,
Hebburn,
Tyne and Wear,
NE31 1EY

Manufacturers of **Suprolastic** fibre fill bitumen solution and other roof protection and repair compounds

London Chemical Co,
Batchworth Island,
Church Street,
Rickmansworth,
Hertfordshire,
WD3 1JQ

Manufacturers of a full range of protection products for roofs including **Silvaseal** light and heat reflective elastomeric roofing compound

Conren Chemicals Ltd,
Silhill House,
2235 Coventry Road,
Birmingham,
B26 3NW

Manufacturers of **Flexiproof** liquid elastomeric roof coating for flat or pitched roofs

Glass Guard Products Ltd,
Units 3/4 Engineers Park,
Sandycroft,
Deeside,
Clwyd,
CH5, 2QD

Manufacturers of **Glassguard** polyester based roofing membrane for new roof covering or for repair work on flat roofs

Advanced Building Products Ltd,
Unit 17C,
Shrub Hill Industrial Estate,
Shrub Hill Road,
Worcester,
WR4 9EL

Manufacturers of **ABP Silverflex,** a solvent-based aluminised elastomeric membrane for waterproofing all types of conventional roof structure. A full advisory service is available

R.F.J. Products,
158 Sherwell Valley Road,
Chelston,
Torquay,
Devon,
TQ2 6ET

Manufacturers of the **Jenny Twin** slate fixer for attaching replacement slates both invisibly and permanently

The National Federation of Roofing Contractors,
15 Soho Square,
London,
W1V 5FB

Publishing a regional register of all specialist roofing companies listing their areas of specialism

Building Centre Group,
26 Store Street,
London,
WC1E 7BT

National centre for the **Building Bookshop,** permanent product displays, and full trade literature for building products of all kinds. Information on regional centres available from this address

Kestner Building Products Ltd,
Station Road,
Greenhithe,
Kent,
DA9 9NG

Manufacturers of glass reinforced plastic **reproduction period rainwater fittings**

Glynwed Foundries,
Sinclair Works,
PO Box 3,
Ketley,
Telford,
Shropshire,
TF1 1BR

Manufacturers of **cast iron rainwater pipes and guttering**

Hunter Building Products Ltd,
Woolwich Industrial Estate,
Nathan Way,
London,
SE28 0AE

Manufacturers of rainwater goods in a variety of colours including the **Highflo** and

Squareflo systems, and the unusual **Leafgo** blockage preventer. Publishers of the free booklet, *How to fit and replace rainwater systems*

Alumasc Ltd,
Burton Latimer,
Kettering,
Northamptonshire,
NN15 5JF

Manufacturers of the **Alumasc** diecast aluminium rainwater systems in modern and traditional designs and in a variety of pre-coloured or plain finishes. Manufacturers of the **Barclay Box** system giving a clean, unbroken eaves line

Wavin Plastics Ltd,
PO Box 12,
Hayes,
Middlesex,
UB3 1EY

Manufacturers of **Osma** uPVC rainwater goods including the **Roundline** and **Squareline** systems. Publishers of free installation booklets relating to each of the Osma systems

Yorkshire Imperial Plastics Ltd,
PO Box 166,
Leeds,
West Yorkshire,
LS1 1RD

Manufacturers of **Rymway** uPVC rainwater systems in several colours which include eaves guttering in various lengths up to 6 metres. Publishers of free installation and product review booklet entitled, *Rymway rainwater, soil, waste, & overflow systems*

Paragon Plastics Ltd,
Broomhouse Lane,
Edlington,
Doncaster,
South Yorkshire,
DN12 1ES

Manufacturers of **uPVC rainwater systems** in roundline and squareline patterns

Key Terrain Ltd,
Aylesford,
Maidstone,
Kent,
ME20 7PJ

Manufacturers of **Terrain** uPVC rainwater systems in roundline and squareline patterns. The range includes outlets for flat roofs and balconies, and pressure release vents for flat roofs. Publisher of free product installation handbook entitled *Rainwater Systems and Roof & Balcony Outlets*

Bartol Plastics Ltd,
Edlington Lane,
Edlington,
Doncaster,
South Yorkshire,
DN12 1BY

Manufacturers of **uPVC rainwater goods** in roundline and squareline patterns. Publishers of free product/installation booklet, *Bartol Rainwater System*

International Seamless Gutters Ltd,
National House,
50–52 Brunel Road,
London,
W3 7XR

Specialists in the on-site manufacture of **roll formed, colour-coated aluminium eaves guttering** in run lengths up to 243m (800ft). Matched with factory made downpipes

Everwarm Cavity Wall Insulation,
Twincliff,
Hillyfields,
Sidcot,
Winscombe,
Nr Bristol
Avon

Manufacturer responsible for the developments of a system of lining faulty eaves gutters such as **concrete channel watercourses.** Waterproofing work undertaken *in situ* using lining materials and polyester resin

Aquaduct System Gutters,
Fairview,
Ten Penny Hill,
Thorrington,
Essex,
CO7 8JB

Manufacturers of the **Aquaduct** seamless aluminium rainwater system with optional aluminium fascia board cover for reducing maintenance work. The system includes gutter grid to prevent leaf blockages

Coburg Brush Ltd,
PO Box 60,
Cheltenham,
Gloucestershire,
GL53 9EZ

Manufacturers of the **Coburg Gutter Grid** which can be fitted to existing eaves gutters to prevent the accumulation of leaves and other wind-borne debris

The Sylglas Co,
Denso House,
Chapel Road,
London,
SE27 0TR

Manufacturers of products including **Sylglas Tape,** a textile fabric impregnated with petroleum compound for weatherproofing rooflights, conservatories, and rainwater goods

Index

References to information contained in the illustrations and photographs are given in *italics*.

abrasive disc machine, 27, 103, 121, 122, 125, 136, 172
abutment, 88, 113, 128, 172, *173, 175,* 177, *177,* 195
access, 16-24, 207
accoustic barriers, 7
advisory services, 204-5
aesthetic appearance, 9, 15
Alumasc, 208, *209,* 211, 213, 217, 249
aluminium:
  flashing, *34*
  nails, 112, 118, 141
  rainwater goods, 208-9, 213, 217
Alutrim, 58, 242
Anderson, 59, 60, 241
Anglia slate, 107, *236,* 244
annular ring-shank nails, 129, *130,* 132
apron flashing, *see* flashing, apron
Aquaseal, 189, 247
Aquatex felt adhesive, 38, 241
asbestos, 31, 106, 117, 164, *196,* 207
asymmetric roofs, 65
atmospheric pollution, *see* pollution
Aviemore, *112*

back gutter, 80, 179, *180,* 181
balconies, 62
balloon, 215, *215*
Bangor, 104, 141
Bardoline shingles, 37, 240
barge boards, *9, 71,* 82, 83, 101, *131,* 144
battens, tiling, 70, 88, 91, 110, 112-13, 114, *114,* 115, *115,* 116, 117, *123, 237*
bench iron, *26*
birds, 91, 96, *119,* 121, 160, 161
birdsmouth joint, *34,* 53, *76,* 77, 78, 79, *81*
Bishore, *61,* 241
bitumen, 43, 58, 60, 168, 216
  backed foil, 169, 185-6, 189, 199, 217, 223, *223, 224*
  boiler, 58, 168
  felt, 37-41, *42, 44,* 45, 46, 59-60, 88, 110-11, *123,* 167,
    see also built-up felt roofing
  impregnated sheeting, 32-3, *33,* 34
  bituminous shingles, 37, 41, 108
Black and Decker jigsaw, 164, 246
boarded roof, 41, 43, 65, 109, 111-12, *112,* 157, 161, *161*
bonnet tiles, 136-7, *137*
bossing, 171-2, *178, 181*
Bradstone, 93, 104, 139, *236,* 243
break iron, *26, 145, 146*
British Board of Agrément, 46, 59
British Lead Manufacturers' Association, 168, 185
British Standards Institution, 7, 29, 240
BS 308, 231
BS 473, 102
BS 550, 102
BS 680, 106
BS 747, 46, 59, 60, 110
BS 1014, 121
BS 4072, 112
BS 4203, 187

BS 4978, 46
BS 5250, 149, 153
  AMD 3025, 161
  AMD 4210, 153, 167
BS 5268, 7, *29,* 65, 67
BS 5472, 30
BS 5534, 94, 109, 112, 125, 140, 143
  AMD 3554, 112
BS 5973, 18
BS 6229, 45
BS 6367, 29
Britorch, 46, 58, 241
broken bond, 133, 138
Building:
  Centre, 205, 248
  Control Dept, 28
  Inspector, 28
  Regulations (England & Wales), 7, 10, 28-9, 30, 32, 37,
    41, 43-4, 46, 47, 77, 80, *94,* 152, 156, 214
  Regulations (Northern Ireland), 28-9
  Research Establishment, 43, 45, 112, 154, 231
  Standards (Scotland) Regulations, 10, 156
built up felt roofing, 46, 56, 58-62
  repair work 188-91
Burton Wire & Tube Co, 23, 239

Caernarvon, 202
Cambrian slate, 107, 113, 243
Cambridge, 96
Cambridgeshire clay, 99
cantilever rafter, 55, 82, 83
capillary action, 88, 101, 138, *139*
capping sheet, 60
carborundum disc cutter, *see* abrasive disc machine
carpenter's square, 77
Catnic, 52, 53, *54,* 55, 241
caulking eaves guttering, 216
Cavity Trays, 36, 178, 179, *179,* 186, 241
ceiling joists, 77, 79, *81,* 94
cement fibre sheeting, *see* fibre cement
  fibre slate, *see* slate
centre gutter, *17*
cheeks of dormer, 88, *169*
chimney:
  pots, 195, 200, *201*
  repairs, 199-201, *203*
  stack, 69, 80-81, 88, 128, 168, 170, 178-81, *180, 181,*
    182, 195, 200
  staging, 23, *23*
chipboard, 38, *54,* 55
chords, 66, *74, 81*
chutes, 25, *25*
Clay Roofing Tile Council, 101, 205, 244
clay tiles, 11, *11,* 93, 95, 96, 99-101, 113, 202, 204
  hand made, 96, 99
  machine made, 99
  pantiles, *101*
  sizes, 99
clinometer, *27*

Index

clips:
  slate, 141
  tile, 93, 118, *118*, *119*, *120*, *135*
cloaked verge tiles, 129, *131*
clout nails, 38-41, 57, 60, 110, 171
coatings, waterproof, 12, 201, *202*
Coburg, gutter grid, 215, 250
cold bonding, 37-41
cold roof, 47, *48*, 49, 56, 60, 156, 167
collar brace, *76*, 77, *94*
colour permanence, 103, 107
comb eaves fillers, *119*, 121
common rafters, 77, 78, *81*, *94*
composite boarding, 55, 56, 168
concrete tiles, 93, 95, 96, 101-3, *102*, 113, 121, 129, 202, 204
condensation, 13, 14, 47, 49, 50, *56*, 129, 153-60, 164, 167, 178, 201, 202
connector plates (trusses), 65, *68*, 69, 70
Consumer's Assoc, 99, 109, 243
Coolag, 50, 55, 168, 241
coping stones, 191, *194*, 203
copper nails, 141, 150, 171, 184
corbelled brickwork, 82, *84*
Cornwall, 93, 104, 201
corrugated iron, *see* zinc coated sheeting
couple roof, 79
Council for Small Industries, 99, 243
cover materials, 46, 87-108, 236
  appearance, 93
  choice, determinants of, 93-6
  costs, 95-6
  function, 91-3
  replacement/maintenance, 191
cover sheets, 25-7, *26*
Cotswolds, 15, 93, 103, 139, 198
Cotswold slate/stone, *92*, 104, 139
counter battens, 50, 111, *112*
creasing slates, 146
cropper, *26*
cross battens, *see* counter battens
Cuprinol, 78
cutting slates/tiles, 121, 136, 145-6, *145*, *146*, 149

damp, 167, 189, 201
decking, 38, 43, 46, 49, *54*, 55, 56, 60, 167, 168
Decra tiles, 107, 244
dentils or dentil slips, 122, *125*
Derbigum, 191, 241
Derbyshire, 204
Devon, 104, 206
dew point, 49, 160
diminishing courses, 104, *105*, 139
Dimos Marketing (UK), 27, 240
disc rivet fixing, 150-51, *151*
DIY, 29, 30, 31, 37-41, 43, 46, 58, 59, 63, 64, 70, 72, *74*, 77, 78, 87, 97, 106, 109, 113, 129, 133, 148, 149, 154, 156, 169, 185, 187, 189, 190, 201, 205
dormer windows, 88, 113, 128, 168, 181
Dorset, 104
double barrier, 88
double lap, 88, *89*, 93, 96, 133-52
downpipes, *see* rainwater downpipes
drainage, 12, 188
drip edging, 38, *39*, *40*, 56, 57, 60
dropped eaves, *17*
dry-fix systems, 113, 129-33, 146-8, *147*, 152, 161
duo pitch, *17*, 65, 67, 69, *76*, 77, 117-52, 166

Duracem slates, 163, 245
Durham, 104

eaves/under-eaves course, 99, 113, 117, 118, 134, 144, 151
eaves detailing, 84, 85, 146, 151
  fillers, *119*, 121
  guttering, *see* guttering
  overhang, 66, 67, *213*
  slates, 138, 140, *141*, 144, 151
  tiles, 87, 99, 113, 114, *114*, 117, 121, 123, *130*, *134*
  ventilation, 161, *161*, 163-5
edging, purpose-made, 58
elastomeric membranes, 190
end lap, *see* head lap
Energy Efficiency Office, 154, *155*, 156, *158*, 246
epoxy resin repairs, 216
Eternit TAC, 31, 245
Europachute (Europalite Ltd), 25, 239
Euroroof, *48*, 49, *61*, 62, 167, 191, 241
Exmouth, 207
expansion gap, 217, 229, *229*
eyebrow dormer, *11*, *17*, 63, 93

fall, 45, 49, 50, 224, 231
fan trussed rafter, 66
fascia, 49, 53, *54*, 55, 84, 85, 110, 113, *114*, 121, 125, *130*, 163, *163*, 165, 211, 224, *225*, *226*; for fixing *see* nailing FEB, 189, 198, 247
Febflex, 216, 247
felt, *see* bitumen felt and sarking felt
Ffestiniog, 204
fibre cement:
  sheeting, 31, 85, 117, 135
  slates, 149-52, 163
  fibreglass insulant, 156, *158*, 159, 161, 167
fillet, cement mortar, 128, 169, 173, 195, 199, *203*
Fink trussed rafter, *65*, 66, 67
fire, 14, 32, 33, 60, 91, 96
firrings, 50, 51
fish scale patterning, *14*, *100*, 101, 105, 107
fixing tiles/slates, 118-51, *see also* nailing slates/tiles
flashing, 58, 128, 168-87, *203*, *see also* bitumen backed foil, lead sheet, soakers and saddle
  apron, *169*, 170, 172, 173-4, *174*, 179, 182, 186
  cover, 174, 177, *178*, *180*, *182*, 187
  PVC, 186-7, *186*
  step, 172, 174-6, *175*, *176*, *180*
flat roofs, *10*, 43-61, 167, 188-91
flat tiles, 113, 174, *176*
flaunching, 195, *200*, 201, *203*
flying gable, 117
frost damage, *11*, 91, 95, 102, 106

gable, 70, *71*, 72, 82-3, 117
gable ladder, 71, 82
gallets, 122, *125*
galvanised steel tiles, 107-8, *108*
Gang-Nail, 69, 72, *74*, 76, 81, 242
gapers, 143
gas vents, 107, 125, 132, 152, 166
gauge, 88, 114-15, 134, 143
geographical location, 93
Glassguard, 190, 248
glass reinforced plastic sheet, 32
Glidevale, 146-8, *147*, 159, 164, *164*, 165, *165*, 245
Glynwed Foundries, 207, 248
grant aid, 13, 159, 205
granular faced tiles, 95, 103

granular fill insulation, *see* vermiculite
groundwork, 88
  hip construction, 123
  valley construction, 125. 126, *128*
gutter brackets, 84, 113, 211, *211*, *226*, *227*
guttering:
  adaptors, 211
  cast iron, 216, 230
  cleaning/maintenance/repair, 216-24
  eaves, 96, 110, 111, 113, *114*, 115, 125, 134, 151, 163, 207-12, 231-5
  fitting, 224-9
  grid, 215, *215*
  plastic, 217
  outlets, 212, *225*, *226*, 231-2
  see also aluminium rainwater goods
Gypglas, 167, 246

hailstones, 91, 102
half bond, 138
handmade clay tiles, *see* clay tiles
hatches, 69
head, 88, 113, *113*, 115
head lap, 60, 88, 103, 113, 114, 115, *120*, 125, 134, 143, 238
herring bone strutting, *54*, 55
high performance roofing, 59, 60, 61
hip:
  board, *74*, 79, 80, 111, *116*, 117
  iron, 122, *123*, *124*
  lead roll cover, 148, 183-4
  mitred, 148, 149, *149*, *184*, 185
  rafter, *see* hip board
  tiles, 122, *123*, *124*, 136-7
hipped roof, *17*, *74*, *75*, 76, 79, 111, 117, 118, 136-7
Hirsh Iron horse & Workgrabber, 23, 239
hoists, 24-5, *25*
hooks, slate, 141, *142*, 150
hopper heads, 207, *212*, 214-15, *214*
Horsham stone, *91*, 93
Hunter Building Products, 208, *208*, 231, 249
Hull, 96
Hydro Air International, 69, 242
Hyparoof membrane, 59, 60, 241

Icovent, *48*, 49, 241
insulation, 31, 47, *48*, 96, 112, 154, 156, 157-8, *157*, *158*, *162*, 168
interlock channel, *see* interlock groove
interlock groove, 88, *89*, 113, 117, 118, *120*, 121, 133
interlocking tiles, 88, *89*, 93, 95, 103, 113, 115, 117
International Truss Plate Association, 69, 242
interstitial condensation, 47, 154
Inverness-shire, *112*
inverted roof, 47, *48*
Isle of Man, 104
Isoflex, 190, 247

jack rafters, *74*, 77, 79, 80
Jenny Twin, 198, *199*, 248
John Caddick, 149, 244
joist hangers, *52*, 72, *73*, 79
joists, 46, 51, *52*, 53, 55, 167

Kent, 99, 102
Kestner Building Products, 207, *207*, 248
Keymer, *236*, 244
kicking slates, 143, 145

king post, 76
knee pads, *26*, 27
Korktaper, 50, 241

ladder access, 20, *20*
Lake District, 95, 104, 105
Lancashire, 104
lap, 72, 88, 111, 114-15, 134, 143
Lead Development Assoc, 168, 169, 173, 185, 246
lead roll cover, 148, 183-4, *184*
  sheet, 168, 170-1
  tools, 171-2, *172*, 181
lean-to roof, *17*, 78, 167, 170, 173, *174*, 186
Leeds, *92*
Leicestershire, 15, 99, 104
Leighton Buzzard, 93
lichens *see* mosses
Llanberis quarry, *147*
loft hatch, 159, *159*
loft space, 154, *155*, 157-9, 161, 163
London Building (Constructional) By-laws, 10, 28-9

mansard roof, 16, 76
margin, 88, 114, 115, 134
Marley products, addresses, 242, 243, 246-7
  advisory service and literature, 85, 103, 205
  dry fix system, 129, *130*, *131*, 132, 161
  guttering, *209*, *210*, 210, 212, 215, 229, 233
  history, 102
  ridge units, *121*, *124*, 125
  special accessories, *119*, 121, *135*
  tiles, *12*, 44, 104, 113, 134, *137*, 139, *236*
  ventilators, *119*, 163, 165, 166
  weatherproofing foils and tapes, *185*, *188*, 189, 246-7
Master Thatchers' Assoc, 99, 243
Measham, 137
mechanical fixing, 118, 122, *124*, 125, 129, *131*, *132*, 152
Mediterranean countries, 88, *89*
Meteorological Office, 93
Metiflash, 36, 241
Metra Non-Ferrous Metals Ltd, 36, 240
mineralised felt, 38-41, 57, 60
mineral wool, 156, 161, 167
mitred hip, *see* hip, mitred
mitred valley, *see* valley, mitred
mono pitch, *17*, 65, 122, 128, 129
Moretonhampstead, 206
mortar mix, 121, 122, 129, 135, *146*, 151, 200, 201
mosses, 104, *193*, 198-9
Multiscaf, 18-20, 239
Mustang Tools, 27, 240

nail and disc rivet system, 150, *151*
nailing slates/tiles, 95, 101, 112, 118, 129, *134*, 135, 139-43, 149-51, *see also* annular, aluminium, copper, zinc
  battens, 115, 117
  fascia boards, 55
nail sickness, 196
National Association of Scaffolding Contractors, 17, 18, 239
National Federation of Roofing Contractors, 205, 248
nibs, 88, 91, 101, 118, 129, 133, 197
nogging, 55
Norfolk, 15, 96, *98*
  reed, 15, 96
Northamptonshire, 101, 104, 138
northern lights, 16
Novolux PVC sheeting, 32-5, *32*, *34*, *35*, 36, 240

Index

oak pegs, 91, 101
ochre, 117, 144
offset, *see* swan's neck offset
OFIC (GB), 33, 240
Onduline, 33-6, 240
open slating, 138, *139*
Osma, 223, 230, 232, 233, 249
outer rafter, 83
outlets, *see* guttering outlets
oversailing, *200*
Oxfordshire, 104

padstone, *74*
pantiles, *89*, 93, 101, 103, 107, 113, 121, 176
parapet, 88, 128, *173*
parrot beak cutter, *145*, 146
particle board, *see* chipboard
PAR timber, 55
patination oil, 148, 171, *171*
pattern rafter, 77
peggies, *see* peg tiles
peg tiles, 90, 91
Pembrokeshire, 93
Penrhyn castle, 104
Penrhyn quarry, 106, *140*, 141, 204, *236*, 245
Permanite, 38, 46, 58, 59, 60, 241
perps, 41, 144, *144*, *145*
Peugeot jigsaw, 164, 246
pigment mortar colouriser, 117, 121
pitch, definition, 44, 45
    determinant of structure, coverings, and weatherproofing,
        34, 38, 85, 87, 93, 95, *110*, 126, *161*, 165, *173*, *174*
    low pitch constraints, 30, 37, 115, 134, 139, 149
    low pitch suitability, 32, *33*, 103, 105, 107, 125, 140, 185
    steep pitch constraints, 113
    steep pitch suitability, 96, 101, 118
pitched roofs, 43, 63-85, 87-151
    repairs, 191-205
pitching up, *42*, 43, 191
plain tile, 88, *89*, 93, 95, 101, 103, 121, 133-8, *134*
    sizes, 99, 134
    undercloak, 117
Planning Office, 28
Planox, 108, 244
Platon system, 108, 244
pollution, 12, 91, 105, 107, 141, 168, 195
polyester membranes, 190-91, *see also* elastomeric
    membranes
polymeric systems, 60
ponding, 45, 188
Portland cement, 103, 121, 122, 135
pour and roll roofing, *44*, 58
pre-felted chipboard, 55
profiled sheeting, 30-36
profiled tile, 88, 95, 103, 107, 113, 117, *119*, 121, 122, 129
protractor rule, 27, 27-8
Purldek, 55, 168, 241
purlin and rafter construction, 76-80
purlins, 34, 36, 38, 63, 64, 65, *76*, 77, 78, *81*, *94*
pvc flashing, 36
pvc sheet, 14, 30-31, *32*, 34-6, *34*, *35*, *36*

queen post, 76
quoins, 80

Rabone Chesterman, 28, 240
rafters, 64, 65, 73, *76*, 77-8, 79, *94*, 113, 156
rainfall intensity, 231

rainwater:
    connectors, 212, *213*
    creep, 88, 115
    discharge, 43, 45, 62, 95, 96, 99, 111, 113, 115, 121,
        134, 135, 146, 179, 188, *198*, 206, 233
    downpipe adaptors, 229
    downpipes, 212-15, 229, 232, 233-5, *see also* outlets
    drainage, 206-35
    system design, 230-35
random slating/tiling, 104, *105*, 139
Readyscaf Chimneydeck, 23, 239
Ready Scaffolding, 24, 239
reclaimed materials, 93, 105, 109, 137, 143-4
reconstituted stone, 103-4
reconstructed stone, *92*, 93, 103-4
re-covering pitched roofs, 202-5
Red Bank, 121, 137, *200*, 201, 244
Redland products, address, 243
    advisory service, 95, 103, 205
    Cambrian slate, 107, 113
    dry fix system, 129, *131*, 132, *132*, 133, *133*
    history, 102
    special accessories, *125*, 128, 177
    tiles, 44, 60, 113, 121, *236*
    ventilators, 110, *128*, 162, 163, 165, 166, *166*
regional style, 15, 93, 97, 103
regulations, *see* Building Regulations
Reigate, 101, 102
Render Stop, 172, 242, 247
repairs on roofs, 188-205
ridge:
    board, *76*, 77, 78, 79, 122, 125, 166, 183
    capping, *34*, 36, 148, *see also* lead roll cover
    course, 99, 114, *114*, 118, 140, *see also* tops tile
    sedge/rushes, 97
    tiles, 100, *100*, 107, 114, *114*, 115, *121*, 122-5, *122*, 128,
        129, *130*, *131*, 132, 137, 149, 152, *193*, *203*
    tree, *see* ridge board
    ventilation, 161-3, 165-7
Roman tiles, *89*, 93, 95, 99, 101, *102*, 103, 113, *113*, 121
roofing square, 27
roof ladder, 18, 21-3
roof lights, 15, 128
Roofpave, 62, 241
roof shape, 93, 113
Rosemary tiles, *236*, 244
Rothwell, *92*
Ruberglas 105 slates, 41, 241
Ruberoid, 38, 59, 241
Rubertherm, *50*, 168, 241
Ryton soffit board, 164, 246

saddle, 181-3, *182*, *183*
safety, 8, 16
sandstone slates, 15, *91*, 93, 139
Sandtoft tiles, *236*, 244
sarking felt, 10, 59, 88, 96, 110-11, *110*, 112, 113, 114,
    117, *118*, 125, *127*, 154, 156, 161, *161*, 163, 228
scaffold, 16, 17, *18*
scarf joint, 85
Scotland, 105, 109, *112*, 139, 140
secret gutters, 181
setting out for tiling-slating, 114-15, 117, 144-5
sheep's teeth, 91
Sheffield, 207
shingles, western red cedar, 108
shoe, 215, 229
Shropshire, *13*, *90*, 91

side lap, 60, *88*
side-lock grooves, *see* interlock groove
side-lock shunt, 89, 117
single lap, *88*, *89*, 93, 96, 113
Skil jigsaw, 164, 246
skew cutting, 80
skew nailing, 70, 78, 115
slate, 138-51, *203*, 238, *see also* cutting slates/tiles
  asbestos, *196*
  clipping, 107
  fixing, 111, 113, 117, 133, 144-5, 150-2
  hooking, 141, *142*
  nailing, head and centre, 139-43
  natural, 88, *89*, 104-6
  plastic concrete, 107
  replacement, 197-8
  sizes, 105
  slate-and-a-half, 146
  stone, *13*, *90*, *92*
  synthetic, 93, 103, 106-7, *106*, 149-52, *150*, *151*, 204
  undercloak, 117
  Welsh, 88, 95, 96, 139-49, *147*, 166, 202, *236*
slater's axe, *143*, 145, *146*, *see also* zax
slater's hammer, *26*, 27
slater's rip, *26*, 27, 166, 197, *197*
slating felt, *see* sarking felt
sloping roofs, 44
soakers, *148*, 149, 176-7, *177*, *178*, 180, 184, 185
soffit, *48*, *54*, *71*, 83, 85, 163, 164
Soffitex eaves ventilators, 164, 246
soil stacks, *128*, 129
Solarbar, 60, 241
solar effects, 12, 13, 60, 61, 91, 189
solar reflective chippings, 188, *190*
spalling, 99, 105, 202, *204*
Spandoboard, 56, 242
Spanish slate, 106
Spanish tiles, *89*
spans, 47, 64, 66, 67, 69, 79, 167
splay cut, 78, 80, 84
sprocketed eaves, *17*, 63
sprocket piece, 85, 110
square four hipped roof, *17*
Staffordshire, 99, 137, 148
stand-off stays, 21, *21*
Stanley tools, 77
stepped flashing, *see* flashing, stepped
Stoke-on-Trent, 149
Stokesay castle, *13*, *90*, 91
stone slate, *see* slate
stop ends, 212, *227*, *228*, 229
strengthening a roof, *94*
stress graded timber, 47, 51, 77
structural integrity, 11, *95*, 107, 113, 129
subtrate, 88
Sumaco, 23, 239
sun decks, 62
Surrey, 101
Sussex, 15, *91*, 93, 99
swan's neck offset, 213, *213*, 230, *230*
Swithland slate, 15, 104
Sylglas, 217, *218*, *221*, 250
synthetic slate, *see* slate

TAC, 31, *236*, 245
tails, 88, 113, *113*
tarpaulins, *see* cover sheets
tenantable repair jobs, 10

terrace roofs, *170*
terminology, 16, *17*, 87-91
thatch, 12, 15, *15*, *88*, 93, 96-8, *97*, *98*, 206
Thatching Advisory Service, 99, 243
thermal insulation, 12, 13, 156, *see also* insulation
thermal movement, 61, 173, 201, 208, 217, 229
tile-and-a-half units, 133, *134*
tile repairs, 197-8
tiles, *see individual types of tile*
tingle, 197, *198*
tongued and grooved boarding, 55, 85, 164
tools, 24-8, 45
tops slates, 138
tops tiles, *124*, 129, 135, *135*
torching, 89
torch-on roofing, 43, *44*, 46, 58
tower access, 18-20, *18*
Town & Country Planning Act, 28
trestles, 23
trimmers, *80*, 81
T roof junction, *74*, 75
trough tile, *89*, 96, 113
truss clips, 70
trussed rafters, 7, *63*, 64-76, 122, 125, 132
  bracing, 67-72, 113
  chimney opening, 81
  erection, 69-72, *74*
  handling, 66-7, *68*
  mono, *74*, 75
  ordering, 66
  storage, 66, *67*
  water tank, 72-3
Tunnel, 31, 245
turf roofs, 108

undercloak, 114, 117-18, *119*, *120*, 135, 144
undereaves, slate/tiles, *see* eaves slates, eaves tiles
underlays, 88, 89, 96, 109-12
Universal Roof Hook, *22*, 23, 239
upstand, 41, 56, 58
uPVC, 129, *131*, 132, 133, 207, 208, 210, 213, 217, 224, 229

valley, 41, 76, 80, 111, 118, 125-7
  groundwork, 126-7, *128*, 137
  gutter, 80, 211, *212*
  laced, 137
  lead lined, 80, 126, *127*, 170, 185
  mitred, 149, 185
  open, 80, 125, 126, *127*
  plastic moulded, 80, 125, 126, *128*
  repairs, 192, *194*, *203*
  swept, 137
  tiles, 80, 137, *138*
  trough tiles, *125*, *126*
vapour, 49, 154, 156, 158
vapour barrier/check, 47, *48*, 49, 55, 56, 156-7, 160, 167, 168
ventilation, 14, 47, *48*, 49, 84, 85, *119*, 129, 152, 153, 160-67, 191, *see also* eaves ventilation and ridge ventilation
ventilators, 49, 84, 85, 107, 110, 125, 128, *131*, 132, 149, 152, 160-67
verge finishes and treatments, 82, *104*, 114, 117, 119, 129, 133, *134*, 135, 146, 151, *152*, *see also* cloaked verge tiles
verge, right/left hand, 87, 118, *119*, *120*, 121
verge tiles, fixing, 121, 151

Index

vermiculite, exfoliated, 156
vermin 49, 91, 119, 121, 160, 161

Wales, 104-6, 183, 201
wall bolts, *78*, 79
wall plate, 52, 53, 64, 70, *76*, 77
warm roof, 47, *48*, 49, 55, 56, 60, 156-7, 167-8
water check, 58
water tanks, 11, 66, 69, 72-3
wbp plywood, 38, 47, *54*, 55, 85, 164
weatherproofing, 168-87
weather resistance, 10, 11, 43, 46, 60, 63, 84, 89, 91, 93, 105, 109, 111, *112*, 115, 118, 128, 129, 134, 138, 168, 173, 178, 185, 189
webs, 66
Welsh slate, *see* slate
welt, 59, 60
West Country, 12

Weston Hyde Products, 33, 34, 240
White Seal Stairways, 207, 239
Wiltshire, 104
wind effect/damage, 10, 11, 46, 63, 91, 93, 105, 109, 111, 115, 118, 129, *131*, *132*, 139, 140, 150, 171, 173, 183, *184*, 186
windows, roof, 113

YIP, 230, 249
Yorkshire, *92*, 93, 104

Zamba slate ventilator, 149, 166, 245
zax, *26*
zinc:
    coated sheeting, 31, 35, 36
    flashing, 12, *34*, 36, 195
    nails, 141